T0271020

Design, Analysis, and Manufacturing of Lightweight Composite Structures

Design, Analysis, and Manufacturing of Lightweight Composite Structures provides a thorough guide to composite materials and their applications, suitable for students of all levels, as well as those in the industry. Covering established theory as well as cutting-edge developments in the field, this book is an essential companion to anyone interested in composite materials.

Discussing the mechanical properties of advanced composites and their materials, this book describes testing and evaluation, focusing on sustainability in manufacturing. Looking at how composite materials can form structural components, this book is centered around how to design and analyze these materials as appropriate to different applications. It discusses micromechanics, stiffness matrices, and numerical calculations using MATLAB®, Excel, and Python. It also covers failure, applied forces, strain, and stress, alongside finite element analysis of composites.

This book is suitable for students and researchers in the field of composites, mechanical design, micromechanics, mechanics of solids, and material science. It also has relevance to the automotive industry.

Design, Analysis, and Manufacturing of Lightweight Composite Structures

Hamid Dalir, Siddharth Bhaganagar, Nicholas Frimas, and Seyedeh Fatemah Nabavi

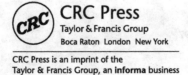

CRC Press
Taylor & Francis Group
Boca Raton London New York

CRC Press is an imprint of the
Taylor & Francis Group, an **informa** business

MATLAB® is a trademark of The MathWorks, Inc. and is used with permission. The MathWorks does not warrant the accuracy of the text or exercises in this book. This book's use or discussion of MATLAB® software or related products does not constitute endorsement or sponsorship by The MathWorks of a particular pedagogical approach or particular use of the MATLAB® software.

First edition published 2024
by CRC Press
2385 NW Executive Center Drive, Suite 320, Boca Raton FL 33431

and by CRC Press
4 Park Square, Milton Park, Abingdon, Oxon, OX14 4RN

CRC Press is an imprint of Taylor & Francis Group, LLC

ISBN: 9781032551401 (hbk)
ISBN: 9781032551418 (pbk)
ISBN: 9781003429197 (ebk)

DOI: 10.1201/9781003429197

Typeset in Times
by codeMantra

Dedicated to the intrepid souls who ceaselessly explore new horizons, never relenting in their quest to forge uncharted paths.

Contents

Preface xi
About the Authors xiii

Chapter 1   Micromechanical Behavior of Lamina 1

    1.1   Some Practical Definitions 1
    1.2   Introduction 4
    1.3   How to Find E_1? 10
    1.4   How to Find E_2? 12
    1.5   How to Find v_{12}? 16
    1.6   How to Find G_{12}? 18
    1.7   Global and Local Coordinates 20
    1.8   Stress and Strain 23
    1.9   Classical Plate Theory 25
    1.10   What Is a Thin Plate? 30
    References 32

Chapter 2   Introduction to ABD Matrix 33

    2.1   Finding Stress and Strain 33
    2.2   Writing a Program to Find *ABD* Matrices 36
    2.3   Hygrothermal Stress/Strain 42
    2.4   Piezoelectric Composites 45
    2.5   Failure Analysis in Composite Materials 45

Chapter 3   Rectangular Composite Beams Being Bended and Loaded
             Axially 48

    3.1   Introduction 48
    3.2   Stress-Strain Relations for On-Axis and Off-Axis
           Composite Elements 48
    3.3   Hooke's Law 53
    3.4   Using the Strength of Materials Approach, Bend
           Curved Beams 55
    3.5   FEM Simulation Models 61
    References 63

Chapter 4   Composite Beams 64

    4.1   Conceptual Design 64
    4.2   Deflection Design 64

4.3   Strength Design 66
4.4   Buckling Design 66
4.5   Behavior of a Column 68
References 69

Chapter 5   Stiffened Panels and Plates 70

5.1   Bending of a Plate 70
5.2   Buckling of a Plate 70
5.3   Stiffened Panels 73
References 81

Chapter 6   Effects on the Environment, Fatigue, and Performance
of Fiber Composites 82

6.1   Fatigue 82
6.2   Fatigue Damage 83
6.3   Empirical Relations for Fatigue Damaged and Fatigue
Life 86
6.4   Impact 87
6.5   Environmental-Interaction Effects 92
6.6   Applicable Problems 95
References 95

Chapter 7   Discontinuous Basalt Fiber-Reinforced Hybrid Composites 98

7.1   Basalt Fibers 98
7.2   Hybrid Composites 101
7.3   Property Prediction 103
7.4   Applications 103
7.5   Thermoplastic Hybrid Composites 104
7.6   Thermoset Hybrid Composites 105
7.7   Basalt Fiber Mat-Reinforced Hybrid Thermosets 106
7.8   Hybrid Fiber Mat-Reinforced Hybrid Thermosets 107
7.9   Conclusion 108
References 108

Chapter 8   Natural Fiber Composites 112

8.1   Introduction 112
8.2   Characteristics 112
8.3   Factors Affecting NFPC's 113
8.4   Applications of NFPC 113
8.5   Natural Fiber Composite (NFC) Design 114
8.6   Material Selection 114

8.7   Concept Design...115
8.8   Incorporating Sustainable Design with Other
       Concurrent Engineering Processes during Product
       Development..116
8.9   Theory of Inventive Problem-Solving (TRIZ).................117
8.10   Applications..120
8.11   Conclusion 122
References 123

Chapter 9   Vibration and Noise...125

9.1   Composite Materials in the Automotive Industry 125
9.2   Understanding of Multilayer Composite Materials..........125
9.3   Case Studies..127
9.4   Conclusion 135
References 135

Chapter 10   Additive Manufacturing in Composites: Fundamentals of
             Processes 136

10.1   What Is AM?..136
10.2   What Are Types of Shaping?..139
10.3   What Are Types of Additive Manufacturing Process?......141
10.4   Extrusion-Based Process...142
10.5   Powder-Based Process...144
10.6   Photopolymerization-Based Process or Vat
       Polymerization (VP)..146
10.7   Comparison of Additive Manufacturing Processes..........147
References 149

Chapter 11   Additive Manufacturing in Composite: Characteristics.............151

11.1   Design for AM Polymer Composites 151
11.2   Mechanical Characteristics 153
11.3   Acoustic Characteristics Driven Design...........................158
11.4   Filler Materials in Additive Manufacturing
       Composites 160
11.5   General Applications for AM in Composites.....................162
11.6   Electrical and Electromagnetic Characteristics 163
11.7   Thermal Conduction and Expansion 165
References 169

Chapter 12 Additive Manufacturing in Composite: Applications
and Models 174

12.1   Introduction 174
12.2   AM Mechanical Applications (2021-VP) 174
12.3   Biomedical Applications 183
References 188

Index... 195

Preface

Welcome to *Design, Analysis, and Manufacturing of Lightweight Composite Structures*, a comprehensive guide that aims to equip readers with a thorough understanding of composite materials and their applications. This book is tailored for individuals of all levels, including students, researchers, manufacturers, and designers working in the dynamic field of composite materials and structures.

In this book, we embark on a journey through every aspect of modern scientific and technological advancements in composite materials. Our exploration encompasses topics such as the physical, chemical, and mechanical properties of advanced composites and the materials constituting them. We delve into theoretical and experimental studies, bridging the gap between microscopic and macroscopic behaviors. This book emphasizes testing and evaluation, with a special focus on environmental effects and reliability. Moreover, we introduce novel techniques for manufacturing various types of composites and shaping structural components using these materials. Through a series of chapters, we explore the design and analysis of composite structures for specific applications. The central purpose of this book is to create a comprehensive "one-stop shop" for all composite-related knowledge. Whether you are a scholar, industry professional, or researcher seeking clarity and insight into composite materials, this book will serve as your guiding light. We commence with an introduction to the concept of composite materials, exploring their significance in various industries and intricacies of production. As we progress, we delve deeper into the micro and macro mechanics of these materials, drawing comparisons with traditional metals.

Throughout this book, we meticulously detail the analysis of composite layups, employing stiffness matrices, transfer matrices, and the ABD matrix. To facilitate numerical calculations, we guide readers in utilizing software tools such as MATLAB®, Excel, and Python. The discussion on failure criteria enhances the reader's comprehension, enabling them to determine strains for applied forces and moments on different composite members.

The beauty of this book lies in its cohesive approach to connect theoretical concepts with practical applications. Readers will witness the seamless integration of software programs discussed in earlier chapters, providing a comprehensive understanding of their functioning and significance in composite analysis.

We strive to empower readers with the ability to perform finite element analysis (FEA) using ABAQUS. Step-by-step instructions and practical examples accompany composite layup creation and analysis, reinforcing the learning experience. The scope of *Design, Analysis, and Manufacturing of Lightweight Composite Structures* is ambitious yet focused. Our goal is to present all pertinent knowledge related to composite structure design and analysis for the industry. Whether you begin with no prior knowledge of composites or are already well versed in the subject, this book is designed to cater to your needs. Starting from the basics and building a solid foundation, this book progresses to advanced theoretical and mathematical concepts, ensuring a comprehensive learning experience.

The authors, drawing from their expertise and experience in both learning and teaching composite materials, have meticulously curated this book with the utmost dedication. Our aim is to inspire curiosity, foster understanding, and pave the way for innovation in the field of composites. We hope that readers find this book a valuable resource, serving as both a guidebook for learning and a dictionary of formulas and methodologies for reference. With enthusiasm and anticipation, we invite you to embark on this educational journey, embracing the world of composite materials and their boundless potential in modern engineering and design.

So, let us begin our exploration into the fascinating realm of composite structures and design!

MATLAB® is a registered trademark of The MathWorks, Inc. For product information, please contact:

The MathWorks, Inc.

3 Apple Hill Drive

Natick, MA 01760-2098 USA

Tel: 508-647-7000

Fax: 508-647-7001

E-mail: info@mathworks.com

Web: www.mathworks.com

About the Authors

Dr. Hamid Dalir is a Professor and Graduate Chair and the Director of the Advanced Composite Structures Engineering Laboratory (ACSEL) at Purdue University, Indianapolis, with faculty appointments in Mechanical Engineering, and Motorsports Engineering. He earned his doctorate in Mechanical Engineering from the Tokyo Institute of Technology, Japan. Prior to joining Purdue, Dalir held the position of the Technical Director of the Syracuse Center of Excellence for Analysis and Design as well as an Associate Professor in the Department of Mechanical and Aerospace Engineering at Syracuse University. He also worked as a Senior Airframe Structures Engineer at Bombardier Aerospace.

Dr. Dalir's research interests are in composite materials and damage mechanics; computational mechanics; multi-disciplinary design optimization; and multi-scale modeling and material characterization. He is a Senior Member of IEEE, AIAA, and ASME. Dalir has co-authored more than 100 peer-reviewed archival publications. His research is supported by NSF, AFRL, IEDC, DOD, Indy Car and CEG, besides several foundations and industries including Dallara, motorsports OEMs, and their suppliers.

Siddharth Bhaganagar is a graduate student in the Advanced Composite Structures Engineering Laboratory (ACSEL) at Purdue University, Indianapolis. He was born and raised in Hyderabad, India. Growing up, he was fascinated with innovation and engineering in motorsports industry, and this interest led to some early exposure in the manufacturing industry where he learnt about composites while interning at the Defense Research and Development Organization.

Nicholas Frimas is a graduate student in the Advanced Composite Structures Engineering Laboratory (ACSEL) at Purdue University, Indianapolis. He was born in Durban, South Africa, and came to the United States for his undergraduate studies in mechanical engineering in 2017. He was given the Plater International Scholarship as well as the Outstanding Technical Scholarship to attend Purdue University and graduated with Honors in 2021.

Seyedeh Fatemeh Nabavi is a graduate student in the Advanced Composite Structures Engineering Laboratory (ACSEL) at Purdue University, Indianapolis. She completed her master's degree in mechanical engineering in 2017. Her thesis won the best thesis award. She has granted/pending national and international patents. Currently, she is focusing on characteristics modeling of AM process under the supervision of Dr. Hamid Dalir at Purdue University.

1 Micromechanical Behavior of Lamina

1.1 SOME PRACTICAL DEFINITIONS

1.1.1 LAMINA

The core of a laminate is an arrangement of woven or unwoven fibers in a matrix that is flat (or occasionally curved). Figure 1.1a and b shows two typical flat laminates as well as the principal axes of the material that are parallel and perpendicular to the direction of the fibers, including laminates with woven fibers and unidirectional fibers. Strong, stiff fibers make up the majority of the principal load-bearing or reinforcing component. The matrix could be biological, metallic, ceramic, or carbon-based. The fibers' needs are met by the matrix, which also acts as a means of load distribution and preservation for the fibers.

The latter function is especially important if a fiber breaks, as depicted in Figure 1.2. The load from one section of a broken fiber is transferred to the matrix, followed by the other section of the damaged fiber and fibers close by at that position. The mechanism for transmitting load is the shearing tension that develops in the matrix and fights against pulling out the split fiber. This load-transfer mechanism is the only method whisker-reinforced composite materials can support any load greater than the natural matrix strength. The properties of the fibers, matrix, and lamina components have only been briefly discussed so far. Figure 1.3 illustrates their stress-strain reaction. Fibers often display linear elastic activity, despite the fact that the behavior of reinforcing steel bars in concrete is more similar to that of elastic-perfectly plastic.

Aluminum, many polymers, and several composite materials exhibit elastic-plastic behavior, which is essentially nonlinear elastic behavior, in the absence of

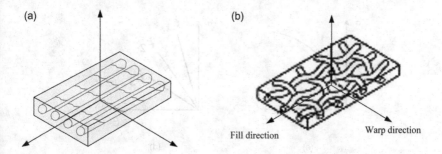

(a)   (b)

Fill direction   Warp direction

FIGURE 1.1 Types of laminate with (a) unidirectional fibers and (b) woven fibers.

DOI: 10.1201/9781003429197-1

1

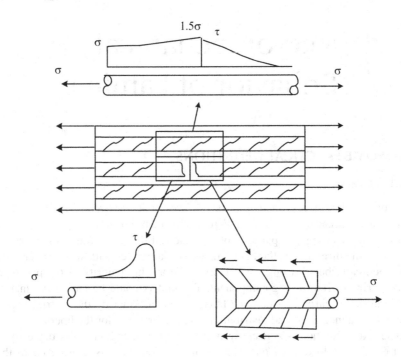

FIGURE 1.2   The broken fiber on matrix.

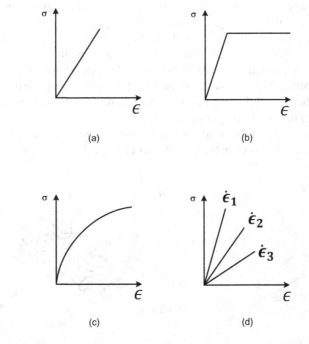

FIGURE 1.3   Stress Strain behaviour, (a) Linear elastic, (b) Elastic-Perfectly plastic, (c) Elastic-Plastic, (d) Viscoelastic $\dot{\epsilon}_1 > \dot{\epsilon}_2 > \dot{\epsilon}_3$.

unloading. Viscoelastic, if not viscoplastic, stress-strain behavior is typically seen in resinous matrix materials. The various stress-strain relations are also referred to as constitutive relations since they describe the mechanical makeup of the material. Fiber-reinforced composite materials like boron-epoxy and graphite-epoxy are usually referred to as linear elastic materials since the majority of the strength and stiffness is given by essentially linear elastic fibers. To improve that approximation, some kind of plasticity, viscoelasticity, or both must be considered (viscoplasticity). Modeling or idealizing the behavior of composite materials in structural design has not gotten much attention.

1.1.2 LAMINATE

A laminate is a bonded stack of laminates with several principal material direction orientations, as shown in Figure 1.4. As can be seen, the layers in the fiber orientation of Figure 1.4 are not symmetrical about the laminate's center surface.

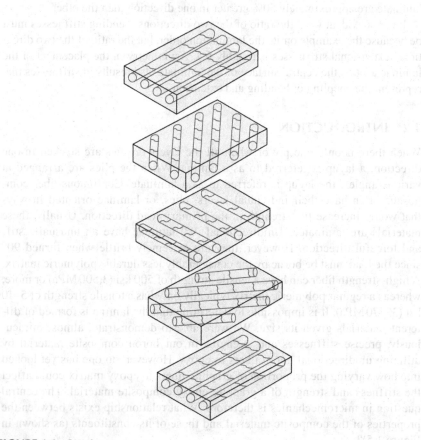

FIGURE 1.4 Laminate.

The layers of a laminate are normally held together by the same matrix material that is used in individual laminates. As a result, a lamina's surfaces are coated with some of the matrix material present in the lamina, which is then used to attach the lamina to the laminae next to it without the usage of additional matrix material. Plates of different materials can be used to create laminates, or in this case, layers of laminates with fiber reinforcement. A laminated circular cylindrical shell can be constructed by winding resin-coated fibers on a moveable core structure known as a mandrel initially with one orientation to the shell axis, then another, and so on, until the required thickness is obtained.

Adjusting the directional dependency of the strength and stiffness of a composite material to meet the loading conditions of the structural element is one of the key objectives of lamination. Laminated structures are particularly suitable for achieving this purpose since the fundamental material orientations of each layer can be changed to suit different requirements. If, for example, six of ten layers of a ten-layer laminate are oriented in one way and the other four are oriented at 90° to that direction, the strength and the extensional stiffness of the resulting laminate are approximately 50% greater in one direction than the other.

It is not evident what the ratio of the two directions' bending stiffnesses must be because the example omits the lamination order, but the ratio of the two directions' extensional stiffnesses is roughly 6:4. Asymmetry in the placement of the laminae around the central surface of the laminate also results in stiffnesses that represent the coupling of bending and extension.

1.2 INTRODUCTION

When there is only one ply or when all the layers or plies are stacked in one direction, a layup is referred to as a lamina. When the plies are arranged at various angles, the layup is referred to as a laminate. Continuous-fiber composites often have their individual layers, plies, or laminae oriented in ways that would increase the strength in the primary load direction. Usually, these materials are laminated. Unidirectional (0°) laminae have an unusually stiff and forceful direction. However, they are extremely brittle when turned 90° since the load must be borne by the considerably less durable polymeric matrix. A high-strength fiber can have a tensile strength of 500 ksi (3,500 MPa) or more, whereas a regular polymeric matrix typically only has a tensile strength of 5–10 ksi (35–70 MPa). It is impossible to determine that the lamina is formed of different materials given its size. We were able to demonstrate, almost miraculously, precise stiffnesses and strengths in our boron composite material by utilizing unidirectional boron fibers in epoxy. However, no one has yet looked into how varying the proportion of graphite fibers to epoxy matrix could affect the stiffness and strength of a graphite-epoxy composite material. The central question in micromechanics is therefore: What relationship exists between the properties of the composite material and those of its constituents (as shown in Figure 1.5)?

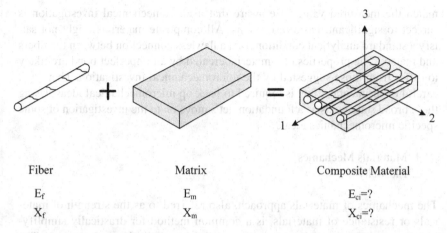

Fiber	Matrix	Composite Material
E_f	E_m	$E_{ci}=?$
X_f	X_m	$X_{ci}=?$

FIGURE 1.5   The basics of micromechanics including fiber, matrix, and composite material.

The optimum method to achieve that stiffness and strength for a composite of two or more materials must be chosen for a solid reason, just like how a certain stiffness or strength of material must be chosen for a specified structural application. In other words, how can the ratios of the component parts be altered to give the composite the necessary stiffness and strength?

By distinguishing between micromechanics and macromechanics, two sub-fields of composite material behavior, it is made easier to divide the aforementioned efforts appropriately:

- **Micromechanics** is the study of the behavior of composite materials, specifically the behavior of heterogeneous composite materials, and includes a detailed examination of the interactions between the constituent elements
- **Macromechanics** is the study of the behavior of composite materials under the premise that they are homogenous, with the effects of the constituent elements only being discernible as the composite material's averaged apparent qualities

The properties of the component materials can therefore be used to theoretically determine a lamina's features, or they can be assessed empirically in the "as produced" condition. To put it another way, lamina qualities can be assessed physically, they can be measured using micromechanics methods, and we can use the features in a micromechanical examination of the structure. Knowing how to anticipate features is necessary when creating composite materials that must have precise visual or macroscopic qualities. As a result, micromechanics is a natural complement to macromechanics when considered in the context of materials design as opposed to a structural analysis environment. Real design power is revealed when the micromechanical predictions of a lamina's properties match the measured values. Be aware that a micromechanical investigation is subject to significant, basic restrictions. All composite materials might not satisfy a standard analytical condition, like a flawless connection between the fibers and matrix. The properties of a material created by an imperfect bond are likely to differ from those suggested by the micromechanical investigation. Therefore, careful experimental study is required to back up micromechanical ideas. Using these broad statements as a foundation, let's move on to the investigation of some specific micromechanics ideas.

1. Materials Mechanics
2. Elasticity

The mechanics of materials approach, also referred to as the strength of materials or resistance of materials, is a common method for drastically simplifying assumptions about the presumptive behavior of mechanical systems. The elasticity approach actually consists of at least three independent approaches: (1) bounding principles, (2) precise solutions, and (3) approximative solutions.

A stronger commitment to the physical laws of equilibrium, deformation continuity and compatibility, and stress-strain relations than in material mechanics distinguishes each method. Both fundamental strategies will be covered in this chapter.

All micromechanics approaches have as their main objective determining the elastic moduli, stiffnesses, or compliances of a composite material in relation to the elastic moduli of the component materials. When calculating the elastic moduli of a fiber-reinforced composite material, for instance, the properties of the fibers and the matrix as well as their respective volumes must be taken into account:

$$C_{ij} = C_{ij}\left(E_f, \upsilon_f, V_f, E_m, \upsilon_m, V_m\right) \tag{1.1}$$

where $C_{ij} = C_{ij}\left(E_f, \upsilon_f, V_f, E_m, \upsilon_m, V_m\right)E_f$ and υ_f are Poisson's ratio and Young's modulus for an isotropic fiber, respectively. The V_f is defined as Eq. (1.2):

$$V_f = \frac{Volume\ of\ fibers}{Total\ volume\ of\ composite\ material} \tag{1.2}$$

The condensed notation for the Hooke's law linking stresses and strains is $\sigma_i = C_{ij}\varepsilon_{ij}$, where $i,j = 1,\ldots,6$.

Where σ_i represents the stress components expressed in the x, y, and z co-ordinates; C_{ij} represents the stiffness matrix; and ε_j represents the strain components.

Micromechanics techniques for composite materials characterization include an additional and supplementary purpose of determining the strengths of the composite material in relation to the strengths of the component materials. For example, when determining the strength of a fiber-reinforced composite material, it is important to consider both the volume (relative to the total volume of the composite material) and the strengths of the fibers and the matrix. Consider the criteria being applicable for the matrix substance. The aforementioned criteria could be modified to take into account changes in strengths under compressive and tensile stresses. The requirements for isotropic fibers and/or matrix materials may also be streamlined. In reality, the functional relationship's structure for composite materials' compression strength might astound us:

$$X_i = X_i \left(X_{if}, V_f, X_{im}, V_m \right) \tag{1.3}$$

$$X_i = X, Y, S = Composite\ material\ strengths \tag{1.4}$$

$$X_{if} = X_f, Y_f, S_f = Fiber\ strengths \tag{1.5}$$

The aforementioned criteria could be modified to take into account changes in strengths under compressive and tensile stresses. The requirements for isotropic fibers or matrix materials may also be streamlined. In reality, the functional relationship's shape for compression strength in composite materials is unexpected.

Little research exists on micromechanical theories of strength. Micromechanical theories of stiffness have been given some thought. We shall concentrate on those stiffness theory elements that most clearly express the aim of micromechanics along with those that are employed most frequently (like the Halpin-Tsai equations). The available strength data will be summarized, much as stiffness theories.

Regardless of the micromechanical stiffness technique used, the fundamental constraints on handling composite materials still apply. The lamina, fibers, and matrix will be described in this case as follows:

- **The lamina** is initially free of tension, linearly elastic, macroscopically homogenous, and macroscopically orthotropic.
- **The fibers** are made of homogenous, isotropic, regularly spaced, linearly elastic, properly aligned, and perfectly bonded material.
- **The matrix** is a material that is void-free, isotropic, homogeneous, and linearly elastic.

In the matrix, between the fibers, or in the matrix itself, there must also be no voids (i.e., the bonds between the fibers and matrix are perfect). While some of these restrictions appear to be extremely doable, others should be quickly ruled out as being rather implausible. For instance, even though the matrix appears to be void-free because of a limited number of voids, the connections between the fibers and the matrix are likely not flawless. The most important concept in discussions is the representative volume element, which is the smallest area or part of a material over which stresses and strains can be regarded as macroscopically uniform while the volume still has the right proportions of fiber and matrix, i.e., it is still representative of the composite material and its constituents by volume. The stresses and strains, however, are not homogenous at the microscopic level because of the heterogeneity of the material. The size of the volume element is essential as a result. There is normally only one fiber in a representative volume element, although more fibers can be required. The representative volume element in a composite lamina comprised of unidirectional fibers has one dimension that is made up of fiber spacing. One of the other two dimensions is lamina thickness or fiber spacing in the thickness direction if the lamina has more than one fiber. The third dimension was randomly selected. A typical representative volume element for a lamina with unidirectional fibers is shown in Figure 1.6.

A lamina with bidirectional fibers requires a representative volume element that is significantly more complex than one with unidirectional fibers. The representative volume element has two dimensions if the weaving geometry is disregarded, which is determined by the matching fiber spacing. The quantity of fibers in the thickness is what ultimately determines the third dimension. If the actual weave geometry (curved fibers) needs to be taken into account, a finite element representation of the representative volume element, as shown in Figure 1.7, may be desirable. Under the supposition that the matrix is perfect, the fiber and the surrounding matrix are both represented by finite elements in the shape of triangles and quadrilaterals, including squares.

No matter what analytic approach is used, the representative volume element must be carefully defined and used. The representative volume element, which is equivalent to the free-body diagram in statics and the dynamics diagram in micromechanics, is actually crucial to the study. The representative volume element has a higher order than the free-body diagram since deformations and stresses are also taken into account in addition to forces.

The results of the micromechanics research of composite materials including unidirectional fibers will be presented as plots of a specific mechanical property vs. the fiber volume percentage. The fiber volume percentage and a number of potential functional correlations between properties are presented schematically

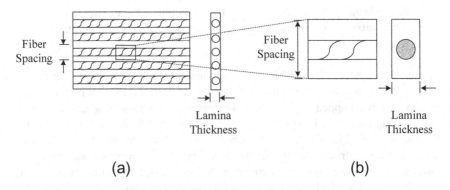

(a)   **(b)**

FIGURE 1.6   The lamina with unidirectional fibers, a representative volume element; (a) macroscopically homogeneous lamina and (b) microscopically heterogeneous representative volume element.

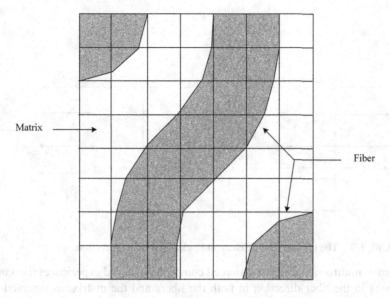

FIGURE 1.7 A representative volume element for a woven lamina is represented by a finite element model.

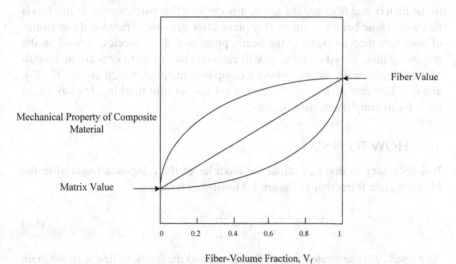

FIGURE 1.8 The results of common micromechanics forms.

in Figure 1.8. These functional correlations' upper and lower bounds will also be established.

The key characteristic of the mechanics of materials approach is the requirement to make some simplifying assumptions about the mechanical behavior of a composite material in order to reach a feasible solution. Every assumption must be tenable, which means that it must have a compelling justification for being true (assumptions in mechanics cannot be incongruous!). The most crucial supposition

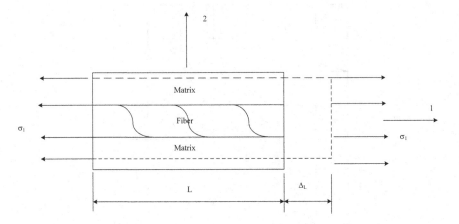

FIGURE 1.9 The typical volume element loaded in the first direction.

is that a unidirectional fiber-reinforced composite material experiences the same stresses in the fiber direction in both the fibers and the matrix, as depicted in Figure 1.9. If the strains were not the same, a fracture between the fibers and the matrix is implied [1]. Therefore, the concept has some merit. Since the strains in the matrix and fiber are the same, it is obvious that parts normal to the 1-axis that were plane before straining stay plane after stressing. Traditional mechanics of materials theories, such as the beam, plate, and shell theories, depend on the preceding idea. Based on this, we will calculate the apparent orthotropic moduli of a unidirectionally fiber-reinforced composite material, which are E_1, E_2, E_{12}, and G_{12}. Remember that any forecast is not just wishful thinking, but has a firm basis for its simplifying assumptions.

1.3 HOW TO FIND E_1?

It is necessary to first determine the modulus of the composite material in the fiber direction (direction 1). Figure 1.9 illustrates how:

$$\varepsilon_1 = \frac{\Delta_L}{L} \tag{1.6}$$

As a result, ε_1 is accurate for both the matrix and the fibers, in line with the main premise. If both component materials behave elastically, the stresses in the fiber direction are therefore equivalent.

$$\sigma_f = E_f \varepsilon_1 \qquad \sigma_m = E_m \varepsilon_1 \tag{1.7}$$

The representative volume element's cross-sectional area A is impacted by the average stress σ_1, σ_f, which has an effect on the fibers' cross-sectional area A_f and the cross-sectional area of the matrix A_m. Consequently, the typical volume element of a composite material experiences a force of:

$$P = \sigma_1 A = \sigma_f A_f + \sigma_m A_m \tag{1.8}$$

On substituting Eq. (1.7) in Eq. (1.8), we get:

$$\sigma_1 = E_1 \varepsilon_1 \tag{1.9}$$

$$E_1 = E_f \frac{A_f}{A} + E_m \frac{A_m}{A} \tag{1.10}$$

However, the matrix and fiber volume fractions can be represented as:

$$V_f = \frac{A_f}{A}, \quad V_m = \frac{A_m}{A} \tag{1.11}$$

Therefore:

$$E_1 = E_m V_m + E_f V_f \tag{1.12}$$

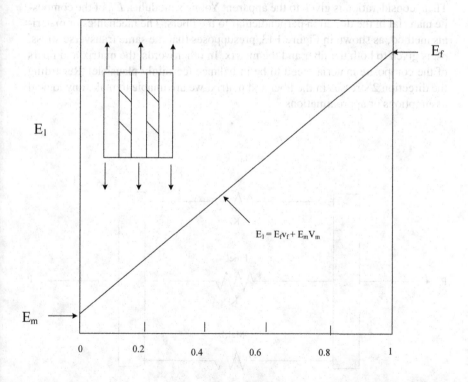

$$E_1 = E_f V_f + E_m V_m$$

Fiber-Volume Fraction, V_f

FIGURE 1.10 Young's modulus in direction 1 (E_1) versus fiber-volume fraction variation.

The apparent Young's modulus of the composite material in the direction of the fibers is depicted graphically in Figure 1.10 according to the rule of mixes. According to the rule of mixtures, the apparent Young's modulus E fluctuates linearly from E_m to E_f as V_f rises from 0 to 1. Typically, the matrix modulus is substantially smaller than the fiber modulus. As a result, the fiber modulus predominates over the composite modulus E_1 for typical practical fiber volume fractions of around 0.6. As long as the fiber volume fraction is not close to zero, even substantial changes in E_m have relatively little impact on E_1 (definitely not proportional to the change in E_m). As a result, we consider E_1 to be a fiber-dominated feature.

A straightforward springs-in-parallel model of the load distribution between the fiber and matrix is shown in Figure 1.11. There, the fiber spring bears the majority of the given load if $k_f \gg k_m$ and all springs deform equally (the equal-strains assumption). Examining the experimental data in relation to the projected straight line in Figure 1.12 will help you understand the practical significance of this study for E_1. The accord is quite good.

1.4 HOW TO FIND E_2?

Then, consideration is given to the apparent Young's modulus, E_2, of the composite material in the direction perpendicular to the fibers. The mechanics of materials method, as shown in Figure 1.13, presupposes that the same transverse stress, σ_2, is given to both the fiber and the matrix. In other words, the matrix and fibers of the composite material need to be in balance (certainly plausible). Regarding the direction 2 stresses in the fiber and matrix, we are unable to make any logical assumptions or approximations.

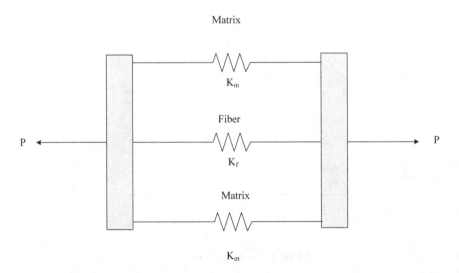

FIGURE 1.11 The shared load in a fiber-reinforced laminate.

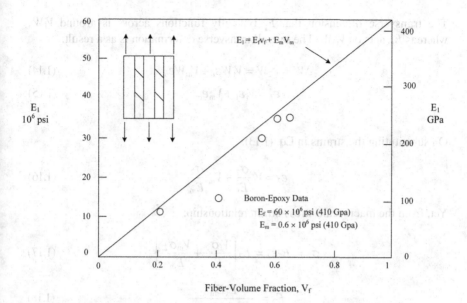

FIGURE 1.12 The measured value of E_1 versus predicted.

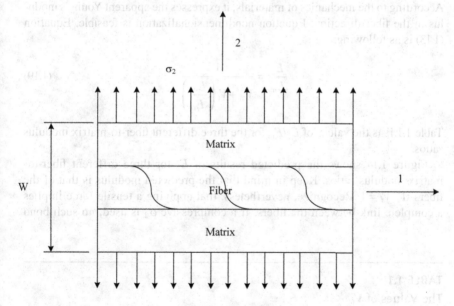

FIGURE 1.13 In the direction 2 loaded sample volume element.

Therefore, the stresses lead to the discovery of the fiber and matrix strains:

$$\varepsilon_f = \frac{\sigma_2}{E_f}, \quad \varepsilon_m = \frac{\sigma_2}{E_m} \tag{1.13}$$

The transverse dimension that E_f typically functions across is around $V_f W$, whereas E_m acts on $V_m W$. The overall transverse deformation is as a result:

$$\Delta W = \varepsilon_2 W = V_f W \varepsilon_f + V_m W \varepsilon_m \tag{1.14}$$

$$\varepsilon_2 = V_f \varepsilon_f + V_m \varepsilon_m \tag{1.15}$$

On substituting the strains in Eq. (1.13):

$$\varepsilon_2 = V_f \frac{\sigma_2}{E_f} + V_m \frac{\sigma_2}{E_m} \tag{1.16}$$

Yet, from the macroscale stress-strain relationship:

$$\sigma_2 = E_2 \varepsilon_2 = E_2 \left[\frac{V_f \sigma_2}{E_f} + \frac{V_m \sigma_2}{E_m} \right] \tag{1.17}$$

$$E_2 = \frac{E_f E_m}{V_m E_f + V_f E_m} \tag{1.18}$$

According to the mechanics of materials, it expresses the apparent Young's modulus in the fiber direction. Equation nondimensionalization is feasible. Equation (1.13) is as following:

$$\frac{E_2}{E_m} = \frac{1}{V_m + V_f \left(\dfrac{E_m}{E_f} \right)} \tag{1.19}$$

Table 1.1 lists the values of E_2/E_m for the three different fiber-to-matrix modulus ratios.

Figure 1.14 shows the predicted results of E_2 for three different fiber-to-matrix-modulus ratios. Keep in mind that the predicted modulus is that of the fibers if $V_f = 1$. Recognize, nevertheless, that applying a tensile force implies a complete link between the fibers. If a compressive σ_2 is used, no such bond

TABLE 1.1
The Values of V_f

| | V_f | | | | | | | |
E_1/E_m	0	0.2	0.4	0.5	0.6	0.8	0.9	1
1	1	1	1	1	1	1	1	1
10	1	1.22	1.56	1.82	2.17	3.56	5.26	10
100	1	1.25	1.66	1.98	2.46	4.80	9.17	100

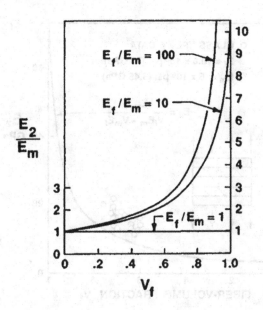

FIGURE 1.14 Young's modulus in direction 2 (E_2) versus fiber-volume fraction variation.

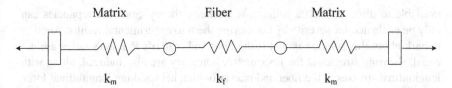

FIGURE 1.15 The sharing of deformation in a fiber-reinforced lamina.

is inferred. Be aware that even if $E_f = 10 \times E_m$!, additional fibers by volume are needed to raise the modulus E_2 to twice that of the modulus of the matrix. That is to say, unless the percentage of fibers is unrealizable high, the fibers do not significantly contribute to the transverse modulus. As a result, the composite material property E_2 is dominated by the matrix. The example volume element loaded in the direction 2 is depicted by a straightforward springs-in-series model in Figure 1.15.

The matrix acts as the supple link in a chain of stiffnesses there. As a result, the matrix's spring stiffness is rather low. On the basis of this, we would anticipate that the matrix deformation dominates the deformation of the composite material.

The presumptions used in the aforementioned derivation are obviously not totally consistent. Equation (1.12) demonstrates that the strain mismatch at the fiber-matrix interface is there. Furthermore, since v_f is not equal to v_m, transversal stresses in the matrix and the fiber being the same is improbable. An alternative is to apply a tight solution to the apparent transverse Young's modulus, which would be made up of precise matching displacements across the fiber-matrix boundary. The only method

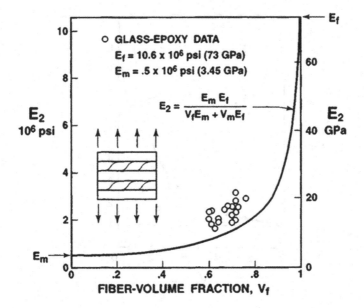

FIGURE 1.16 Predicted vs measured.

available to discover such a solution is elasticity theory. Such discrepancies can only be evaluated for severity by comparing them to experimental results. Another remark about this solution is that if the fiber and matrix's Poisson's ratios are not equal, shearing stresses at the fiber-matrix boundary are also induced, along with longitudinal stresses in the fiber and matrix (with a net resulting longitudinal force of zero). These shearing pressures are inescapably going to emerge under specific stress conditions. Therefore, it is impossible to consider this material characteristic to be undesirable or suggestive of an inappropriate solution.

Figure 1.16 displays the observed values of E_2 together with the estimates of E_2 from Eq. (1.18). There, it is evident that this strategy undervalues the flexible matrix material's contribution to E_2.

1.5 HOW TO FIND v_{12}?

The analysis of E_1 may be used to generate the so-called Poisson's ratio, v_{12}. The primary ratio of Poisson's is:

$$v_{12} = -\frac{\varepsilon_2}{\varepsilon_1} \tag{1.20}$$

when all other stresses are zero and the stress state $\sigma_1 = \sigma$. Following that, Figure 1.17's typical volume element shows the deformations. There, as in the approach to E_1, the basic simplifying premise means that, in the fiber direction, the matrix strains and fiber stresses are equal. Lateral deformation is:

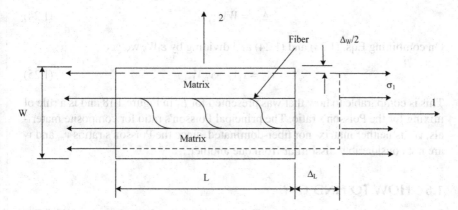

FIGURE 1.17 Typically loaded in the first direction was the volume element.

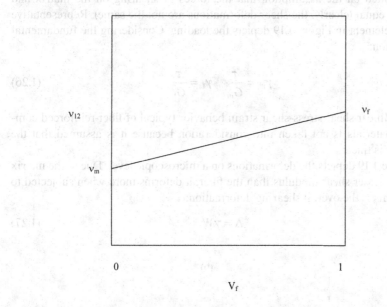

FIGURE 1.18 v_{12}, variant with the fiber-volume fraction.

$$\Delta_w = -W\varepsilon_2 = Wv_{12}\varepsilon_1 \tag{1.21}$$

Which is also:

$$\Delta_w = \Delta_{mw} + \Delta_{fw} \tag{1.22}$$

Transverse deformations Δ_{mw} and Δ_{fw} are somewhat similar to the calculations for the transverse Young's modulus, E_2:

$$\Delta_{mw} = WV_m v_m \varepsilon_1 \tag{1.23}$$

$$\Delta_{fw} = WV_f v_f \varepsilon_1 \qquad (1.24)$$

On combining Eqs. (1.23) and (1.24) and dividing by $\varepsilon_1 W$, we get:

$$v_{12} = v_f V_f + v_m V_f \qquad (1.25)$$

This is comparable to how that was presented for E_1 in Figure 1.18 and is a rule of mixing for the Poisson's ratio. The principal Poisson's ratio for composite materials, v_{12}, is neither matrix- nor fiber-dominated, since the Poisson's ratios v_m and v_f are not considerably dissimilar from one another.

1.6   HOW TO FIND G_{12}?

In the mechanics of materials method, a lamina's in-plane shear modulus, G_{12}, is calculated on the assumption that the forces of shearing on the matrix and fiber are equal (clearly, the shear deformations are not the same). Representative volume element in Figure 1.19 depicts the loading. Considering the fundamental assumption:

$$\gamma_m = \frac{\tau}{G_m} \quad \gamma_f = \frac{\tau}{G_f} \qquad (1.26)$$

The nonlinear shear stress-shear strain behavior typical of fiber-reinforced composite materials is not taken into consideration because it is assumed that the behavior is linear.

Figure 1.19 depicts the deformations on a microscopic level. Due to the matrix having a lesser shear modulus than the fiber, it deforms more when subjected to shear. This is the overall shearing deformation:

$$\Delta = \gamma W \qquad (1.27)$$

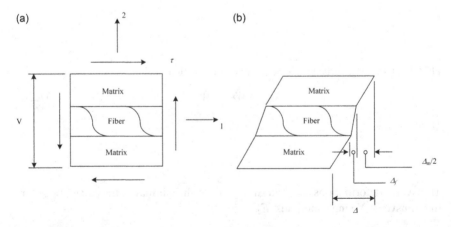

FIGURE 1.19   The shear-loaded representative volume element: (a) shear stress loading and (b) shear deformation.

and consists of microscopic deformations:

$$\Delta_m = V_m W \gamma_m, \quad \Delta_f = V_f W \gamma \tag{1.28}$$

Then, $\Delta = \Delta_m + \Delta_f$, divided by W yields:

$$\gamma = V_m \gamma_m + V_t \gamma_t \tag{1.29}$$

Upon substituting Eq. (1.26):

$$\Delta_m = V_m W \gamma_m, \quad \Delta_f = V_f W \gamma_f \tag{1.30}$$

$$\gamma = \frac{\tau}{G_{12}} \tag{1.31}$$

Equation (1.31) can also be expressed as:

$$\frac{\tau}{G_{12}} = V_m \frac{\tau}{G_m} + V_f \frac{\tau}{G_f} \tag{1.32}$$

Then:

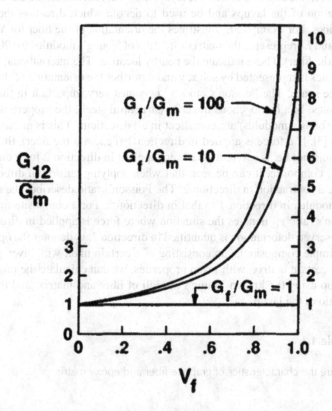

FIGURE 1.20 G_{12}, variant with the fiber-volume fraction.

$$G_{12} = \frac{G_m G_f}{V_m G_f + V_f G_m} \tag{1.33}$$

This was found for the transverse Young's modulus, E_2, and is a similar kind of expression. The equation for G_{12}, like that for E_2, can be normalized by a matrix-related modulus, i.e.,

$$\frac{G_{12}}{G_m} = \frac{1}{V_m + V_f\left(\dfrac{G_m}{G_f}\right)} \tag{1.34}$$

This is displayed in Figure 1.20 as a plot for various G_f/G_m values... rises beyond twice G_m even though $G_f/G_m = 10!$ only for the volume of the fiber bigger than 50% of the whole volume. The shear modulus of composite materials, G_{12}, is matrix-dominated, similar to E_2. The relationship between the measured and anticipated values of G_{12} is similar to that of E_2 in Figure 1.20 [2].

1.7 GLOBAL AND LOCAL COORDINATES

When analyzing composites, a useful tool to use is the concept of local and global coordinates. In the case of a composite layup, the local coordinates would describe the orientation of the layups and be used to denote which direction the property describes. For example, E_1 illustrates the orientation of the fiber for Young's modulus, and E_2 represents the matrix direction of Young's modulus (or 900 with respect to the fiber). These indicate the nearby location. The international system of coordinates is represented by sub x, y and describes the orientation of the laminate in free space. The Poisson's ratio 'v' becomes very important in the study of composites. In a homogenous material (structural steel), the properties of the material (Young's modulus) are consistent in all directions. This is not true for a composite [3]. If a force is applied in direction 1 (i.e., with the fiber), the deformation would not be as severe as if it were applied in direction 2 for a unidirectional fiber composite. It can be seen that when applying tension in direction 1, we can see a contraction in direction 2. The Poisson's ratio describes the ratio of Young's modulus in direction 1 to that in direction 2. For a composite material, the Poisson's ratio v_{12} denotes the situation where force is applied in direction 1 and the observed deformation is quantified in direction 2. v_2 denotes the opposite.

For a simple composite layup consisting of a certain fiber, with given properties, and a certain matrix, with given properties, we can calculate the properties of the layup using the known Young's moduli of fiber and matrix, and the fiber volume ratio of the layup.

Example 1.1

These are the characteristics of graphite fiber and epoxy matrix:

$$E_f = 2.756 \times 10^5 \, \text{MPa}, \quad \upsilon_f = 0.2$$

$$E_m = 2.756 \times 10^3 \, \text{MPa}, \quad \upsilon_n = 0.33$$

Find E_1, E_2 for a graphite/epoxy composite lamina with a fiber volume ratio of 60%?
Find the Poisson's ratios υ_1, υ_2?
Find the shear modulus G_{12}?

Solution

$$E_1 = E_f V_f + E_m V_m$$

$$E_2 = \frac{E_f E_m}{V_m E_f + V_f E_m}$$

$$G_{12} = \frac{G_f G_m}{V_m G_f + V_f G_m}$$

$$V_{12} = v_m V_m + v_f V_f$$

$$V_m = 1 - V_f$$

$$V_m = 1 - V_f$$

$$1 - 0.6 = 0.4$$

$$E_1 = E_f V_f + E_m V_m$$

$$E_1 = (2.76 * 105 * 0.6) + (2.6 * 103 * 0.4)$$

$$E_1 = 1.67 * 105 \, \text{MPa}$$

$$E_2 = \frac{E_f E_m}{V_m E_f + V_f E_m}$$

$$E_2 = 6.78 * 103 \, \text{MPa}$$

$$V_{12} = (v_m V_m - v_f V_f)$$

$$(0.3 * 0.4) + (0.2 * 0.6) = 0.252$$

$$V_{21} = \frac{E_2}{E_1} * V_{12}$$

$$6.78 * \frac{103}{1.67} * 105 * 0.252 = 0.01$$

$$G_f = \frac{E_f}{2(1 + v_f)}$$

$$2.75 * \frac{105}{2(1 + 0.2)} = 1.148 * 105 \, \text{MPa}$$

$$G_m = \frac{E_m}{2(1 + v_m)}$$

$$G_m = 1.0361 * 103 \, \text{MPa}$$

$$G_{12} = \frac{G_m G_f}{V_m G_f + V_f G_m}$$

$$G_{12} = 2.556 * 103 \, \text{MPa}$$

1.7.1 TRANSFORMATION MATRIX

Often when dealing with local and global coordinates, it is beneficial to be able to move from one to the other. We achieve this through the use of the transformation matrix. What is meant by this is that, by using the transformation matrix and inverse transfer matrix, we can move between stress in the local and global coordinate system and vice versa [4].

The transformation matrix $[T]$ is used to convert global to local coordinates ($x \rightarrow 1$):

$$[T] = \begin{bmatrix} \cos^2 \theta & \sin^2 \theta & 2\sin\theta\cos\theta \\ \sin^2 \theta & \cos^2 \theta & -2\sin\theta\cos\theta \\ -\sin\theta\cos\theta & \sin\theta\cos\theta & \cos^2 \theta - \sin^2 \theta \end{bmatrix} \quad (1.35)$$

The inverse transformation matrix $[T]^{-1}$ is used to convert local to global coordinates ($1 \rightarrow x$):

$$[T^{-1}] = \begin{bmatrix} \cos^2 \theta & \sin^2 \theta & -2\sin\theta\cos\theta \\ \sin^2 \theta & \cos^2 \theta & 2\sin\theta\cos\theta \\ \cos\theta\sin\theta & -\cos\theta\sin\theta & \cos^2 \theta - \sin^2 \theta \end{bmatrix} \quad (1.36)$$

For instance, this is how the transformation matrix could be used:

$$\begin{matrix} \sigma_x \\ \sigma_y \\ \tau_{xy} \end{matrix} = \begin{bmatrix} T^{-1} \end{bmatrix} \begin{matrix} \sigma_1 \\ \sigma_2 \\ \tau_{12} \end{matrix} \tag{1.37}$$

1.8 STRESS AND STRAIN

In order to introduce the concept of stress and strain, we must introduce the stiffness matrix. The stiffness matrix, denoted by $[Q]$, is used to calculate stress from strain in the local coordinate system. The inverse stiffness matrix $[Q]^{-1}$ is used to calculate the strain from the stress in the local coordinate system. For global coordinates, we must use the $[\bar{Q}]$ to calculate stress from strain and $[\bar{Q}]^{-1}$ to find strain from stress. The stiffness matrices are symmetric.

Q Matrix: Local

$$\begin{matrix} \sigma_1 \\ \sigma_2 \\ \gamma_{12} \end{matrix} = \begin{matrix} Q_{11} & Q_{12} & 0 \\ Q_{12} & Q_{22} & 0 \\ 0 & 0 & Q_{11} \end{matrix} \begin{bmatrix} \epsilon_1 \\ \epsilon_2 \\ \gamma_{12} \end{bmatrix} \tag{1.38}$$

$$Q_{11} = \frac{E_1}{1 - \gamma_{12}\gamma_{21}} \tag{1.39}$$

$$Q_{12} = \frac{\gamma_{12}E_2}{1 - \gamma_{12}\gamma_{21}} \tag{1.40}$$

$$Q_{22} = \frac{E_2}{1 - \gamma_{12}\gamma_{21}} \tag{1.41}$$

$$Q_{66} = \frac{G_{12}}{E} \tag{1.42}$$

$Q_{12} = \dfrac{\gamma_{12}E_2}{1 - \gamma_{12}\gamma_{21}} Q_{22} = \dfrac{E_2}{1 - \gamma_{12}\gamma_{21}} Q_{66} = \dfrac{G_{12}}{E}$ $[Q^{-1}]$ Matrix: Local:

$\sigma_1 \rightarrow \varepsilon_1 \Rightarrow$

$$\begin{matrix} \varepsilon_1 \\ \varepsilon_2 \\ \gamma_{12} \end{matrix} = \begin{matrix} \dfrac{1}{E_1} & \dfrac{-\gamma_{21}}{E_2} & 0 \\ \dfrac{-\gamma_{12}}{E_1} & \dfrac{1}{E_2} & 0 \\ 0 & 0 & \dfrac{1}{G_{12}} \end{matrix} \begin{bmatrix} \sigma_1 \\ \sigma_2 \\ \tau_{12} \end{bmatrix} \tag{1.43}$$

Q^{-1} Matrix: (Global)

$\sigma_x \rightarrow \varepsilon_x \Rightarrow$

$$
\begin{array}{c}
\sigma_x \\
\sigma_y \\
\tau_{xy}
\end{array}
=
\begin{bmatrix}
\overline{Q}_{11} & \overline{Q}_{12} & \overline{Q}_{16} \\
\overline{Q}_{12} & \overline{Q}_{22} & \overline{Q}_{26} \\
\overline{Q}_{16} & \overline{Q}_{26} & \overline{Q}_{66}
\end{bmatrix}
\begin{array}{c}
\varepsilon_x \\
\varepsilon_y \\
\gamma_{xy}
\end{array}
\tag{1.44}
$$

$$\overline{Q}_{11} = Q_{11}\cos^4\theta + 2(Q_{12} + 2Q_{66})\sin^2\theta\cos^2\theta + Q_{22}\sin^4\theta \tag{1.45}$$

$$\overline{Q}_{12} = (Q_{11} + Q_{22} - 4Q_{66})\sin^2\theta\cos^2\theta + Q_{12}(\sin^4\theta + \cos^4\theta) \tag{1.46}$$

$$\overline{Q}_{22} = Q_{11}\sin^4\theta + 2(Q_{12} + Q_{66})\sin^2\theta\cos^2\theta + Q_{22}\cos^4\theta \tag{1.47}$$

$$\overline{Q}_{16} = (Q_{11} - Q_{12} - 2Q_{66})\sin\theta\cos^2\theta + (Q_{12} - Q_{22} + Q_{66})\sin^3\theta\cos\theta \tag{1.48}$$

$$\overline{Q}_{26} = (Q_{11} - Q_{12} - 2Q_{66})\sin^3\theta\cos\theta + (Q_{12} - Q_{22} + Q_{66})\sin\theta\cos^3\theta \tag{1.49}$$

$$\overline{Q}_{66} = (Q_{11} + Q_{22} - 2Q_{12} - 2Q_{66})\sin^2\theta\cos^2\theta + Q_{66}(\sin^4\theta + \cos^4\theta) \tag{1.50}$$

$\overline{Q_{11}Q_{12}Q_{22}}$ $\overline{Q_{16}Q_{26}Q_{66}}$ Q^{-1} Matrix: (Global)

$\sigma_x \rightarrow \varepsilon_x \Rightarrow$

$$
\begin{array}{c}
\varepsilon_x \\
\varepsilon_y \\
\gamma_{xy}
\end{array}
=
\begin{bmatrix}
\overline{S}_{11} & \overline{S}_{12} & \overline{S}_{16} \\
\overline{S}_{12} & \overline{S}_{22} & \overline{S}_{26} \\
\overline{S}_{16} & \overline{S}_{26} & \overline{S}_{66}
\end{bmatrix}
\begin{array}{c}
\sigma_x \\
\sigma_y \\
\tau_{xy}
\end{array}
\tag{1.51}
$$

$$\overline{S}_{11} = S_{11}\cos^4\theta + (2S_{12} + S_{66})\sin^2\theta\cos^2\theta + S_{22}\sin^4\theta \tag{1.52}$$

$$\overline{S}_{12} = (S_{11} + S_{22} - S_{66})\sin^2\theta\cos^2\theta + S_{12}(\sin^4\theta + \cos^4\theta) \tag{1.53}$$

$$\overline{S}_{22} = S_{11}\sin^4\theta + 2(S_{12} + S_{66})\sin^2\theta\cos^2\theta + S_{22}\cos^4\theta \tag{1.54}$$

$$\overline{S}_{16} = (2S_{11} - 2S_{12} - S_{66})\sin\theta\cos^3\theta - (2S_{22} - 2S_{12} - S_{66})\sin^3\theta\cos\theta \tag{1.55}$$

$$\overline{S}_{26} = (2S_{11} - 2S_{12} - S_{66})\sin^3\theta\cos\theta - (2S_{22} - 2S_{12} - S_{66})\sin\theta\cos^3\theta \tag{1.56}$$

$$\overline{S}_{66} = 2(2S_{11} + 2S_{22} - 4S_{12} - S_{66})\sin^2\theta\cos^2\theta + S_{66}(\sin^4\theta + \cos^4\theta) \tag{1.57}$$

1.9 CLASSICAL PLATE THEORY

1.9.1 WHAT IS CLASSICAL LAMINATED PLATE THEORY?

The Kirchhoff-Love-proposed classical plate theory for homogeneous, isotropic materials is directly extended by the classical laminate theory (CLT). A number of relationships involving stress, force, moments, deformation, and strain for laminated plates are found using this theory.

1.9.1.1 Why We Need Classical Laminated Plate Theory?

CLPT makes it possible to move quickly from a lamina, a fundamental building block, to the intended outcome, a laminated structure. Finding effective and reasonably accurate simplification assumptions is the entire process that enables us to shift our focus from a challenging, three-dimensional elasticity problem to a manageable, two-dimensional mechanics of deformable bodies problem.

1.9.2 CLASSICAL PLATE THEORY (CPT)

The classical plate theory (CPT), also known as Kirchhoff-Love theory, which expresses the stress-strain relations of a thin plate with homogenous, isotropic materials, is necessary to understand in order to achieve CLPT. These relationships will then be expanded to include thin laminated plates.

A thin-plate application of the Euler-Bernoulli beam theory is the Kirchhoff-Love theory (CPT). Using Kirchhoff's hypotheses as a foundation, Love created the theory in 1888. It is believed that the three-dimensional plate can be represented in two dimensions using a mid-surface plane.

In this hypothesis, the following kinematic presumptions are made:

1. Following deformation, lines that are normal to the mid-surface stay straight.
2. Even after deformation, straight lines that are normal to the mid-surface stay that way.
3. A deformation does not alter the plate's thickness.

This section discusses the classical lamination theory's stress and deformation hypotheses, which are based on the mechanics of materials. Using this method, we are able to progress continuously from the structural laminate, the basic building block, to the final result, the lamina. The entire process is designed to find practical and comforting simplifying assumptions that enable us to change our attention from a difficult three-dimensional elasticity problem to a manageable two-dimensional mechanics of deformable bodies problem.

Due to the stress and deformation assumptions that are a fundamental part of classical lamination theory, the terms "classical thin lamination theory" and even "classical laminated plate theory" would be more accurate. We will keep using the term "classical lamination theory" despite the fact that it is a useful simplification of the complex language. In the composite materials literature, the phrase "classical lamination theory" (CLT) is frequently abbreviated.

1.9.3 STRESS-STRAIN OF A LAMINA

The stress-strain relationships in primary material coordinates for a lamina of an orthotropic material under plane stress are:

$$\begin{bmatrix} \sigma_1 \\ \sigma_2 \\ \gamma_{12} \end{bmatrix} = \begin{bmatrix} Q_{11} & Q_{12} & 0 \\ Q_{12} & Q_{22} & 0 \\ 0 & 0 & Q_{66} \end{bmatrix} \begin{bmatrix} \varepsilon_1 \\ \varepsilon_2 \\ \gamma_{12} \end{bmatrix} \tag{1.58}$$

The engineering constants used in the equation define the lowered stiffnesses, Q_{ij}. Any other coordinate system results in stresses in the plane of the lamina that are:

$$\begin{bmatrix} \sigma_x \\ \sigma_y \\ \gamma_{xy} \end{bmatrix} = \begin{bmatrix} \bar{Q}_{11} & \bar{Q}_{12} & \bar{Q}_{16} \\ \bar{Q}_{12} & \bar{Q}_{22} & \bar{Q}_{26} \\ \bar{Q}_{16} & \bar{Q}_{26} & \bar{Q}_{66} \end{bmatrix} \begin{bmatrix} \varepsilon_x \\ \varepsilon_y \\ \gamma_{xy} \end{bmatrix} \tag{1.59}$$

where the reduced stiffnesses, Q_{ij}, and the transformed reduced stiffnesses, $\bar{Q}_{ij}$, are provided.

Due to the arbitrary orientation of the constituent laminae, the stress-strain relations in arbitrary in-plane coordinates, or Eq. (1.59), are helpful in defining the laminate stiffnesses. For the kth layer of a multilayered laminate, Eqs. (1.58) and (1.59) may both be conceived of as stress-strain relations. Equation (1.59) may be expressed as:

$$\{\sigma\}_k = [\bar{Q}]_k \{\sigma\}_k \tag{1.60}$$

The next section will discuss the definition of strain and stress variations along a laminate's thickness. The forces and moments on a laminate will then be produced by integrating the stress-strain relations for each layer through the laminate thickness subject to the variations in stress and strain determined in Section 1.3.2.

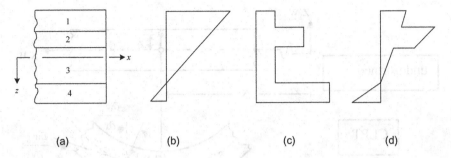

(a) (b) (c) (d)

FIGURE 1.21 Variation of strain and stress; (a) laminate, (b) strain distribution, (c) characteristic moduli, and (d) stress distribution.

1.9.4 VARIATION OF STRESS AND STRAIN IN A LAMINATE

The determination of a laminate's extensional and bending stiffnesses depends on understanding the fluctuation of stress and strain through the laminate thickness. It is assumed that the laminate is made up of perfectly glued laminae. The bonds are also assumed to be non-shear deformable and infinitesimally thin. In order to prevent any lamina from moving in relation to another, the dispositions are continuous across lamina borders. As a result, the laminate functions as a single layer with unique characteristics, which, as we shall see in a moment, forms a structural element. In light of this, if the laminate is thin, it is presumed that a line that was at first straight and perpendicular to the middle surface of the laminate, or a normal to the middle surface, will remain straight and perpendicular to the middle surface even after the laminate is bent, extended, contracted, sheared, or twisted. Z is the middle surface normal's direction in Figure 1.21. Note that γ_{xz} and γ_{yz} are the angles created by the distorted middle surface and a deformed normal. The shearing strains in planes perpendicular to the middle surface must be ignored in order for the normal to the middle surface to remain straight and normal under deformation. Since it is believed that the normals have constant lengths, the strain perpendicular to the center surface is likewise excluded, which results in $\varepsilon_z = 0$. The aforementioned set of hypotheses regarding the behavior of the single layer that functions as the laminate represent the Kirchhoff hypothesis for plates and the Kirchhoff-Love hypothesis for shells (and are the two-dimensional equivalents of the common one-dimensional beam theory assumption that plane sections, i.e., sections normal to the beam axis, remain plane after bending, so the physical justification of the group of assumptions should be obvious). The fact that laminates don't have to be flat should be highlighted; they can really be curved or shell-like. To ascertain the effects of the Kirchhoff theory on the laminate dispositions u, v, and w in the x-, y-, and z-directions, the laminate cross-section in the x–z plane shown in Figure 1.22 is employed. Point B shifts

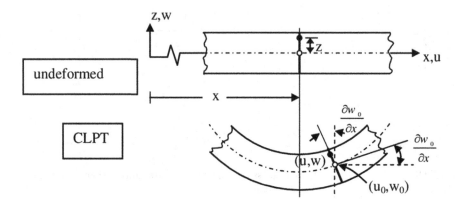

FIGURE 1.22   Transverse normal deformation in accordance with the shear deformation theory.

the middle surface that is warped in the x-direction from the middle surface that is not. The term "middle-surface values of a variable" is given by the symbol "nought" (0). Line ABCD stays straight despite the laminate's distortion; hence, the displacement at point C is:

$$u_c = u_0 - z_c \beta \qquad (1.61)$$

However, because line ABCD continues to be perpendicular to the middle surface despite deformation, β is the slope of the laminate's middle surface in the x-direction, or:

$$\beta = \frac{\partial W_0}{\partial x} \qquad (1.62)$$

The displacement, u, through the laminate thickness at any point z is thus:

$$u = u_0 - z \frac{\partial W_0}{\partial x} \qquad (1.63)$$

Using the Kirchhoff theory, the laminate strains have been reduced to ε_x, ε_y, γ_{xy}. Consequently, $\varepsilon_z = \gamma_{xy} = \gamma_{yz} = 0$. The remaining strains are described in terms of displacements for minimal strains (linear elasticity) as:

$$\varepsilon_x = \frac{\partial u}{\partial X}$$

$$\varepsilon_y = \frac{\partial V}{\partial y}$$

$$\gamma_{xy} = \frac{\partial u}{\partial y} + \frac{\partial V}{\partial X}$$

(1.64)

The stresses for the computed displacements u and v in Eqs. (1.63) and (1.64) are as a result:

$$\varepsilon_x = \frac{\partial u_0}{\partial X} - z \frac{\partial^2 W_0}{\partial x^2}$$

$$\varepsilon_y = \frac{\partial V_0}{\partial Y} - z \frac{\partial^2 W_0}{\partial y^2}$$

(1.65)

$$\gamma_{xy} = \frac{\partial u_0}{\partial y} + \frac{\partial V_0}{\partial X} - 2z \frac{\partial^2 W_0}{\partial x \partial y}$$

$$\begin{bmatrix} \varepsilon_x \\ \varepsilon_y \\ \gamma_{xy} \end{bmatrix} = \begin{bmatrix} \varepsilon_x^0 \\ \varepsilon_y^0 \\ \gamma_{xy}^0 \end{bmatrix} + z \begin{bmatrix} k_x \\ k_y \\ k_{xy} \end{bmatrix}$$

(1.66)

Where the strains in the middle surface are:

$$\begin{bmatrix} \varepsilon_x^0 \\ \varepsilon_y^0 \\ \gamma_{xy}^0 \end{bmatrix} = \begin{bmatrix} \dfrac{\partial u_0}{\partial X} \\ \dfrac{\partial V_0}{\partial Y} \\ \dfrac{\partial u_0}{\partial y} + \dfrac{\partial V_0}{\partial X} \end{bmatrix}$$

(1.67)

Curvatures at the middle surface are:

$$\begin{bmatrix} k_x \\ k_y \\ k_{xy} \end{bmatrix} = \begin{bmatrix} \dfrac{\partial^2 W_0}{\partial x^2} \\ \dfrac{\partial^2 W_0}{\partial y^2} \\ \dfrac{2 \partial^2 W_0}{\partial x \partial y} \end{bmatrix}$$

(1.68)

(The final element in Eq. (1.68) represents the twist curvature of the middle surface.) No superscripts are required for k_x, k_y, and k_{xy} since we only use the center surface's curvatures as a reference point and not any other surface. Because the stresses in Eq. (1.66) take the shape of a straight line, i.e., $y = mx + b$, it is simple to demonstrate that the Kirchhoff theory predicts a linear change in strain along the laminate thickness. The previous strain analysis only applies to plates because of the interactions between strain and displacement in Eq. (1.64). Other shells have more complex strain-displacement relationships; as a result, the Kirchhoff hypothesis for circular cylindrical shells has been simply demonstrated to predict a linear change in strain along the laminate thickness. The previous strain analysis only applies to plates because of the interactions between strain and displacement in Eq. (1.64). For cylinder-shaped shells, since certain shells have more complex strain-displacement relationships, the ε_y component in Eq. (1.64) must be augmented by $\dfrac{w_0}{r}$ where r is the shell radius. The stresses in the kth layer may be represented in terms of the laminate middle-surface strains and curvatures by substituting Eq. (1.66) for the strain variation across thickness in the stress-strain equation (1.60):

$$
\begin{bmatrix} \sigma_x \\ \sigma_y \\ \tau_{xy} \end{bmatrix}_k = \begin{bmatrix} \bar{Q}_{11} & \bar{Q}_{12} & \bar{Q}_{16} \\ \bar{Q}_{12} & \bar{Q}_{22} & \bar{Q}_{26} \\ \bar{Q}_{16} & \bar{Q}_{26} & \bar{Q}_{66} \end{bmatrix}_k \left(\begin{bmatrix} \varepsilon_x^0 \\ \varepsilon_y^0 \\ \gamma_{xy}^0 \end{bmatrix} + z \begin{bmatrix} k_x \\ k_y \\ k_{xy} \end{bmatrix} \right) \tag{1.69}
$$

Even if the strain variation is linear, the $\bar{Q}_{ij}$ might vary for each layer of the laminate, making the stress variation over the thickness of the laminate not always linear. Figure 1.21 illustrates typical strain and stress fluctuations instead, where the stresses are piecewise linear (i.e., linear in each layer, but discontinuous at boundaries between laminae).

Note that after deformation, the center surface's normal stays parallel to it, which is equal to ignoring shearing strains on planes perpendicular to it, that is, $\gamma_{xz} = \gamma_{yz} = 0$ [5].

In addition, the normal is presumed to have constant length, so the strains perpendicular to the middle surface are ignored, and hence, $\varepsilon_z = 0$. Therefore, the plane strain condition can be considered.

1.10 WHAT IS A THIN PLATE?

Plates may be categorized into four varieties, including extremely thin, moderately thin, thick, and very thick, as illustrated in portion B of Figure 1.23,

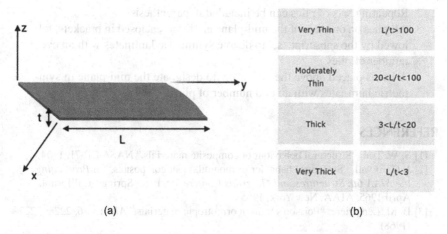

FIGURE 1.23 Thin plate; (a) schematic and (b) types of plates based on their thickness.

TABLE 1.2
Stacking Sequence Notation

Laminate	Description	Layers
[45/30/0]	One layer each of 45°, 30° and 0°	3
45°, 30°	One layer of 30° and −30°	2
[45₂]	Two layers of 45°	2
[(30)₂/0]	Two layers of 30°, one layer of 0°	3
[45/30]ₛ	Symmetric with 45° and 30° layers	4
[±45/±30]ₛ	Symmetric with +45°, −45° and +30° and −30° layers	8
[(±45)₂/(±30)₂]ₛ	Symmetric with two groups of +45°, −45° and two groups of +30°, −30° layer	16
[±θ]ₛ	Symmetric with one layer of +θ and one layer of −θ	4

assuming "t" denotes the plate's thickness and "L" represents a sample length or breadth measurement.

Very thin and moderately thin plates are both covered under the "traditional" idea of plates [6].

The laminates' structures, which have been explained in previous sections, are gathered in stacking form in Table 1.2. The rules, which are related to these laminates, are provided as follows:

1. Each full laminate is contained in a set of brackets.

2. Repeating sets of plies can be included in parenthesis.
3. All plies on one side of the mid-plane are listed enclosed in brackets, followed by the subscript "s," to denote symmetric laminates with an even number of plies.
4. A bar is placed above the center ply to designate the mid-plane in symmetric laminates with an odd number of plies.

REFERENCES

[1] S. W. Tsai, "Structural behaviour of composite materials," NASA CR-71, 1964.

[2] J. C. Ekvall, "Structural behavior of monofilament composites," in *Proceedings of the AIAA 6th Structures and Materials Conference,* Palm Springs, California, 5–7 April 1965, AIAA, New York, 1965.

[3] B. M. Lempriere, "Poisson's ratio in orthotropic materials," *AIAA J.*, 6, 2226–2227 (1968).

[4] R. M. Jones, "Stiffness of orthotropic materials and laminated fiber-reinforced composites," *AIAA J.*, 12, 112–114 (1974).

[5] R. M. Jones, *Mechanics of Composite Materials*, Second Edition, CRC Press, Philadelphia, USA, 1999.

[6] A. Elmarakbi, *Advanced Composite Materials for Automotive Applications*, First Edition, Wiley.

2 Introduction to ABD Matrix

2.1 FINDING STRESS AND STRAIN

The concept of determining total strain at any arbitrary surface, at some distance (z) from the midplane, becomes important when analyzing composites—due to the fact that we can then find the stresses from the strain as outlined in Chapter 1. The total strain anywhere in the laminate can be represented as follows:

$$\varepsilon_x = \frac{\delta U_0}{\delta x} - z\frac{\delta^2 w_0}{\delta x^2} \tag{2.1}$$

$$\varepsilon_y = \frac{\delta V_0}{\delta x} - z\frac{\delta^2 w_0}{\delta y^2} \tag{2.2}$$

$$\gamma_{xy} = \frac{\partial U_0}{\partial y} + \frac{\partial V_0}{\partial x} - 2z\frac{\partial^2 W_0}{\partial x \partial y} \tag{2.3}$$

The above three equations are the total strains for any arbitrary surface distance z from the middle surface.

$$\begin{bmatrix} \varepsilon_x \\ \varepsilon_y \\ \gamma_{xy} \end{bmatrix} = \begin{bmatrix} \varepsilon_x^0 \\ \varepsilon_y^0 \\ \gamma_{xy}^0 \end{bmatrix} + z\begin{bmatrix} k_x \\ k_y \\ k_{xy} \end{bmatrix} \tag{2.4}$$

$$\begin{bmatrix} k_x \\ k_y \\ k_{xy} \end{bmatrix} = -\begin{bmatrix} \partial^2 W_0 / \partial x^2 \\ \partial^2 W_0 / \partial y^2 \\ 2\partial^2 W_0 / \partial x \partial y \end{bmatrix} \tag{2.5}$$

DOI: 10.1201/9781003429197-2

$$
\begin{bmatrix} \varepsilon_x^0 \\ \varepsilon_y^0 \\ \gamma_{xy}^0 \end{bmatrix} = \begin{bmatrix} \dfrac{\partial U_0}{\partial x} \\ \dfrac{\partial V_0}{\partial y} \\ \dfrac{\partial U_0}{\partial y} + \dfrac{\partial V_0}{\partial x} \end{bmatrix} \tag{2.6}
$$

We can see that the strain at an arbitrary surface is comprised of strain at the midplane and curvature at the midplane. The values for strain and curvature at the midplane can be found using the displacements u, v, and w at the midplane, which are in the x, y, and z directions, respectively. These strain and curvature values can also be used to find the moment and forces applied to the laminate. This, however, also means that from the moments and forces, we can find the strain and therefore the stress in the laminate.

Let us first look at finding the moment and forces from strain and curvature. In order to do this, we must introduce the concept of the *ABD* matrices. The [*A*] matrix is the extensional stiffness matrix, [*B*] matrix is the coupling stiffness matrix, and [*D*] matrix is the bending stiffness matrix. The [*B*] matrix will always be 0 if the laminate is symmetric about the midplane.

To find the *ABD* matrices, we must first find the *Q* matrix, as outlined in Chapter 1, for each layer of the laminate using the layers E_1, E_2, G_{12}, and v_{12}. Each layer will have a unique *Q* matrix based on the properties of its fibers and matrix, as well as their relationship to one another. This can be shown in Chapter 1 when calculating E_1, E_2, and Poisson's ratio.

From each layer's *Q* matrix, we must now find the $\bar{Q}$ matrix using the fiber angle (Θ) with respect to the global equations outlined in Chapter 1:

$$
\overline{Q_{11}} = Q_{11}\cos^4\theta + 2(Q_{12} + 2Q_{66})\sin^2\theta\cos^2\theta + Q_{22}\sin^4\theta \tag{2.7}
$$

$$
\overline{Q_{12}} = (Q_{11} + Q_{22} - 4Q_{66})\sin^2\theta\cos^2\theta + Q_{12}(\sin^4\theta + \cos^4\theta) \tag{2.8}
$$

$$
\overline{Q_{22}} = (Q_{11} - Q_{12} - 2Q_{66})\sin\theta\cos^2\theta + (Q_{12} - Q_{22} + 2Q_{66})\sin^3\theta\cos\theta \tag{2.9}
$$

$$
\overline{Q_{16}} = (Q_{11} - Q_{12} - 2Q_{66})\sin\theta\cos^2\theta + (Q_{12} - Q_{22} + 2Q_{66})\sin^3\theta\cos\theta \tag{2.10}
$$

$$
\overline{Q_{26}} = (Q_{11} - Q_{12} - 2Q_{66})\sin^3\theta\cos\theta + (Q_{12} - Q_{22} + 2Q_{66})\sin\theta\cos^3\theta \tag{2.11}
$$

$$
\overline{Q_{66}} = (Q_{11} + Q_{22} - 2Q_{12} - 2Q_{66})\sin^2\theta\cos^2\theta + Q_{66}(\sin^4\theta + \cos^4\theta) \tag{2.12}
$$

Now that we have the $\bar{Q}$ matrices, we can use these to find the *ABD* matrices. Extensional stiffness matrix is as following:

$$
A_{ij} = \sum_{k=1}^{N} \left(\overline{Q_{ij}}\right)_k (Z_k - Z_{k-1}) \tag{2.13}
$$

$$B_{ij} = \frac{1}{2} \sum_{k=1}^{N} \left(\overline{Q}_{ij} \right)_k \left(Z_k^2 - Z_{k-1}^2 \right) \tag{2.14}$$

$$D_{ij} = \frac{1}{3} \sum_{k=1}^{N} \left(\overline{Q}_{ij} \right)_k \left(Z_k^3 - Z_{k-1}^3 \right) \tag{2.15}$$

$$\begin{bmatrix} N_x \\ N_y \\ N_{xy} \end{bmatrix} = \begin{bmatrix} A_{11} & A_{12} & A_{16} \\ A_{12} & A_{22} & A_{26} \\ A_{16} & A_{26} & A_{66} \end{bmatrix} \begin{bmatrix} \varepsilon_x^0 \\ \varepsilon_y^0 \\ \gamma_{xy}^0 \end{bmatrix} + \begin{bmatrix} B_{11} & B_{12} & B_{16} \\ B_{12} & B_{22} & B_{26} \\ B_{16} & B_{26} & B_{66} \end{bmatrix} \begin{bmatrix} k_x \\ k_y \\ k_{xy} \end{bmatrix} \tag{2.16}$$

$$\begin{bmatrix} M_x \\ M_y \\ M_{xy} \end{bmatrix} = \begin{bmatrix} B_{11} & B_{12} & B_{16} \\ B_{12} & B_{22} & B_{26} \\ B_{16} & B_{26} & B_{66} \end{bmatrix} \begin{bmatrix} \varepsilon_x^0 \\ \varepsilon_y^0 \\ \gamma_{xy}^0 \end{bmatrix} + \begin{bmatrix} D_{11} & D_{12} & D_{16} \\ D_{12} & D_{22} & D_{26} \\ D_{16} & D_{26} & D_{66} \end{bmatrix} \begin{bmatrix} k_x \\ k_y \\ k_{xy} \end{bmatrix} \tag{2.17}$$

From these equations, it is apparent that we need a value for Z_k and Z_{k-1}. These values represent the different layers and their distance from the center line in the laminate. For instance, if we start by taking the bottom-most layer to be our start and we know the thickness of each layer, then by drawing out this laminate, as shown in Figure 2.1, we can label the top and bottom of each layer as follows (assuming that each layer is 0.5 mm thick).

Make the midplane 0 and label the distance from the midplane at the edge of each layer. Therefore, we can see that for:

$$\text{Layer 1}: Z_k - Z_{k-1} = -0.5 - (-1) \tag{2.18}$$

$$\text{Layer 2}: Z_k - Z_{k-1} = 0 - (-0.5) \tag{2.19}$$

$$\text{Layer 3}: Z_k - Z_{k-1} = 0.5 - 0 \tag{2.20}$$

$$\text{Layer 4}: Z_k - Z_{k-1} = 1 - 0.5 \tag{2.21}$$

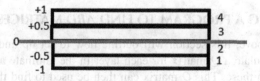

FIGURE 2.1 Laminate thickness.

Now to calculate the A matrix for instance:

$$A_{11} = \left[\bar{Q}_{11}(\text{layer 1})*(-0.5-(-1))\right]+\left[\bar{Q}_{11}(\text{layer 2})*(0-(-0.5))\right]$$
$$+\left[\bar{Q}_{11}(\text{layer 3})*(0.5-0)\right]+\left[\bar{Q}_{11}(\text{layer 4})*(1-0.5)\right] \tag{2.22}$$

$$A_{12} = \left[\bar{Q}_{12}(\text{layer 1})*(-0.5-(-1))\right]+\left[\bar{Q}_{12}(\text{layer 2})*(0-(-0.5))\right]$$
$$+\left[\bar{Q}_{12}(\text{layer 3})*(0.5-0)\right]+\left[\bar{Q}_{12}(\text{layer 4 })*(1-0.5)\right] \tag{2.23}$$

Repeat these steps for all values in the A matrix and then repeat for the B and D matrices, using their respective formulae.

Once we have the ABD matrices, we are able to calculate moment and force from strain. Another useful calculation is to use the inverse ABD matrices to find the strain at any layer using force and moment. To do this, we must find the inverse of the A, B, and D matrices, respectively, and then use the values of the matrices to do the calculation. The inverse can be found easily using MATLAB® or Excel functions. They are then used as follows:

$$\begin{bmatrix} \varepsilon_x^0 \\ \varepsilon_y^0 \\ \gamma_{xy}^0 \end{bmatrix} = \begin{bmatrix} a_{11} \\ a_{12} \\ a_{16} \end{bmatrix} \begin{bmatrix} \hat{N} \\ b \end{bmatrix} \tag{2.24}$$

$$\begin{bmatrix} k_x \\ k_y \\ k_{xy} \end{bmatrix} = \begin{bmatrix} d_{11} \\ d_{12} \\ d_{16} \end{bmatrix} \begin{bmatrix} \widehat{M_y} \\ b \end{bmatrix} \tag{2.25}$$

$$\begin{bmatrix} \varepsilon_x \\ \varepsilon_y \\ \gamma_{xy} \end{bmatrix} = \begin{bmatrix} \varepsilon_x^0 \\ \varepsilon_y^0 \\ \gamma_{xy}^0 \end{bmatrix} + z \begin{bmatrix} k_x \\ k_y \\ k_{xy} \end{bmatrix} \tag{2.26}$$

Z is the distance from the midplane to the plane at which you would like to find the strain. It is highly recommended that one creates a program in MATLAB or Excel to calculate the Q, $\bar{Q}$, ABD, and abd (inverse of ABD) matrices from the properties of each layer.

2.2   WRITING A PROGRAM TO FIND *ABD* MATRICES

As discussed above, this section will outline how to set up and use an Excel program to calculate a Q matrix for each layer in the laminate and then create a $\bar{Q}$ matrix from these. This $\bar{Q}$ matrix can then be used to find the A, B, and D

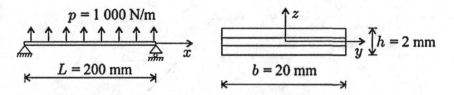

FIGURE 2.2 The dimension of the composite beam.

matrices that represent the response of the layup and can be used to find the stress and strain from the input force and moment.

- **Home Work #1:**

   The following composite beam is made of carbon/epoxy and its layup is (90/0$_2$/90). The layers have equal thickness. Find the maximum deflection of the beam and the slope at supports and the top ply stresses and strains (as shown in Figure 2.2):

$$E_1 = 140\,\text{GPa}, \quad E_2 = 38\,\text{GPa}, \quad G_{12} = 10\,\text{GPa}, \quad \nu_{12} = 0.28$$

Equations for finding the Max deflection of a beam:

The EI terms must be replaced with $\dfrac{b}{d_{11}}$, which we will get from the inverse B and D matrices denoted as b and d. For this example of a simply supported beam, we will use:

$$\delta_{max} = \frac{pl^3}{48\left(\dfrac{b}{d_{11}}\right)}$$

Finding *ABD* matrices: Here we will set up our input region for each layer.

According to Figure 2.3a, the formula will find the Q matrix for each layer, starting with layer 1: Q_{66} is equivalent to the value of G_{12} for layer 1 (as shown in Figure 2.3b) and then layer 2, etc. All other values are calculated from the color-coded equations. For instance, Q_{11} cell holds the equation in green for layer 1 ONLY. Layer 2's Q_{11} would have the values in the same column but moved one row down (i.e., the values associated with layer 2). Instead of $M4/1 - (O4 * P4)$ as used for Q_{11} (layer 1), we would use $M5/1 - (O5 * P5)$ for Q_{11} (layer 2).

As shown in Figure 2.4, the Q matrices should be positioned alongside the "setup" region of the program. The steps in Figure 2.3b are repeated for each layer to get a Q matrix for each one:

In Figure 2.5, a single $\bar{Q}$ matrix (on the right) is found for EACH of our Q matrices. This figure uses color to show the equations used to find each cell of the $\bar{Q}$ matrix. These equations use the values of the Q matrix to find $\bar{Q}$.

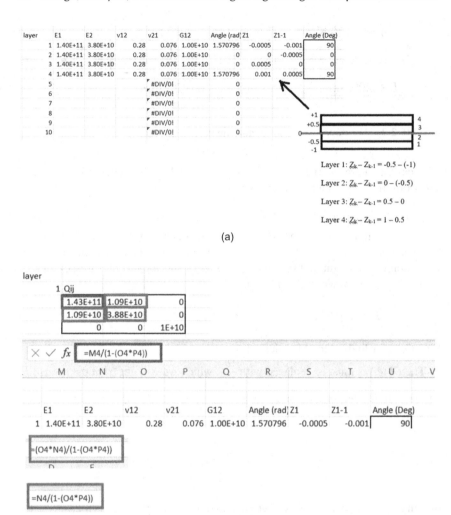

(a)

(b)

FIGURE 2.3 The process of finding Q matrices; (a) setup and (b) equations for Q matrix.

In Figure 2.6, the equation for finding A_{11} (Eq. 2.13) is shown again. In the thick box, it is shown which cells in the $\overline{Q}$ matrices are used in this equation. Similarly, A_{22} would be the exact same equation (eq for A) except $\overline{Q}_{11}$ (layer 1) would be replaced with $\overline{Q}_{22}$ (layer 1). B and D matrices are found in the same way as the A matrix, except we use the formulas:

$$B_{ij} = \frac{1}{2} \sum_{k=1}^{N} \left(\overline{Q}_{ij} \right)_k \left(Z_k^2 - Z_{k-1}^2 \right) \tag{2.27}$$

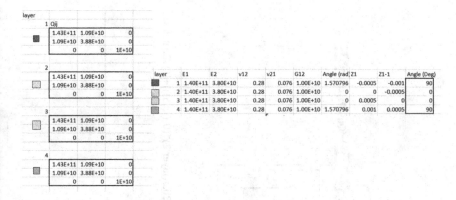

FIGURE 2.4 Showing which layer in setup is associated with each Q matrix.

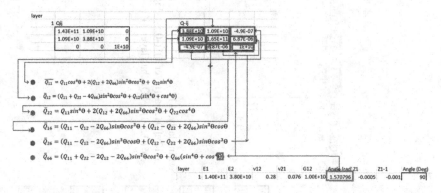

FIGURE 2.5 Finding $\bar{Q}$ matrix from Q matrix.

$$D_{ij} = \frac{1}{3} \sum_{k=1}^{N} \left(\bar{Q}_{ij} \right)_k \left(Z_k^3 - Z_{k-1}^3 \right) \tag{2.28}$$

where Q_{ij} denotes the respective cell in the $\bar{Q}$ matrix (i.e., Q_{12}, Q_{22}). Finding *ABD*:

Use the {=minverse()} command on *A* to get *a*, *B* to get *b*, and *D* to get *d*. Using *ABD* to find strain, stress, and curvature (Figure 2.7):

We determine the moments and forces acting on the beam based on the supports and input loading:

$$M_x = 0 \tag{2.29}$$

$$M_y = \frac{Wl^3}{8} = \frac{(1,000)(0.2)}{8} \tag{2.30}$$

$$M_{xy} = 0 \tag{2.31}$$

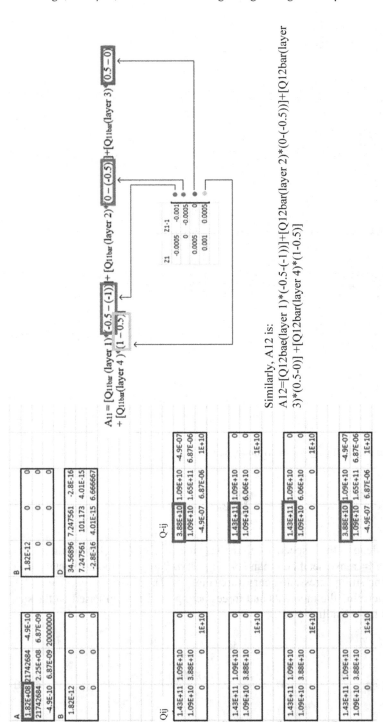

FIGURE 2.6 Finding A_{11}.

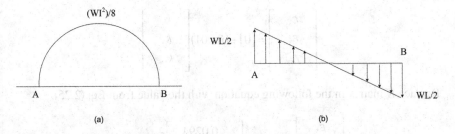

FIGURE 2.7 Balance diagrams; (a) moment and (b) forces.

$$N_x = 0 \tag{2.32}$$

$$N_y = 0 \tag{2.33}$$

$$N_{xy} = 0 \tag{2.34}$$

$$\begin{bmatrix} \varepsilon_x^0 \\ \varepsilon_y^0 \\ \gamma_{xy}^0 \end{bmatrix} = \begin{bmatrix} a_{11} \\ a_{12} \\ a_{16} \end{bmatrix} \begin{bmatrix} \dfrac{\hat{N}}{b} \end{bmatrix}$$

$N=0$. Therefore $\begin{bmatrix} \varepsilon_x^0 \\ \varepsilon_y^0 \\ \gamma_{xy}^0 \end{bmatrix} = 0$

$$\begin{bmatrix} k_x \\ k_y \\ k_{xy} \end{bmatrix} = \begin{bmatrix} d_{11} \\ d_{12} \\ d_{16} \end{bmatrix} \begin{bmatrix} \dfrac{\widehat{M_y}}{b} \end{bmatrix}$$

$$\begin{bmatrix} k_x \\ k_y \\ k_{xy} \end{bmatrix} = \begin{bmatrix} 0.0294 \\ -0.0021 \\ 0 \end{bmatrix} \begin{bmatrix} \dfrac{(100*0.2)/8}{b} \end{bmatrix}$$

$$\begin{bmatrix} \varepsilon_x \\ \varepsilon_y \\ \gamma_{xy} \end{bmatrix} = \begin{bmatrix} \varepsilon_x^0 \\ \varepsilon_y^0 \\ \gamma_{xy}^0 \end{bmatrix} + z \begin{bmatrix} k_x \\ k_y \\ k_{xy} \end{bmatrix}$$

$$\begin{bmatrix} \varepsilon_x \\ \varepsilon_y \\ \gamma_{xy} \end{bmatrix} = [0] + (0.001) \begin{bmatrix} k_x \\ k_y \\ k_{xy} \end{bmatrix}$$

Replace K matrix in the following equation with the value from Eq. (2.25):

$$\begin{bmatrix} \varepsilon_x \\ \varepsilon_y \\ \gamma_{xy} \end{bmatrix} = (0.001) \begin{bmatrix} 0.0294 \\ - \ 0.0021 \\ 0 \end{bmatrix} \begin{bmatrix} \dfrac{25\,\mathrm{Nm}}{0.02} \end{bmatrix}$$

$$\begin{bmatrix} \varepsilon_x \\ \varepsilon_y \\ \gamma_{xy} \end{bmatrix} = (0.001) \begin{bmatrix} 7.35 \\ -5.25 \times 10^{-4} \\ 0 \end{bmatrix}$$

$$\begin{bmatrix} \varepsilon_x \\ \varepsilon_y \\ \gamma_{xy} \end{bmatrix} = \begin{bmatrix} 7.35 \times 10^{-3} \\ 0.0021 \\ 0 \end{bmatrix}$$

Find stress at a layer by using the layer's $\bar{Q}$ matrix:

$$\begin{bmatrix} \sigma_x \\ \sigma_y \\ \tau_{xy} \end{bmatrix} = \begin{bmatrix} \bar{Q}_{11} & \bar{Q}_{12} & \bar{Q}_{16} \\ \bar{Q}_{12} & \bar{Q}_{22} & \bar{Q}_{26} \\ \bar{Q}_{16} & \bar{Q}_{26} & \bar{Q}_{66} \end{bmatrix} \begin{bmatrix} \varepsilon_x \\ \varepsilon_y \\ \gamma_{xy} \end{bmatrix}$$

On solving:

$$\begin{bmatrix} \sigma_x \\ \sigma_y \\ \tau_{xy} \end{bmatrix} = \begin{bmatrix} 2.73 \times 10^8 \\ -6.205 \times 10^8 \\ -3.6 \times 10^8 \end{bmatrix}$$

2.3 HYGROTHERMAL STRESS/STRAIN

In engineering, materials are always exposed to high or low temperatures and moisture. These parameters can engender mechanical strains in a laminated composite and subsequently different kinds of stresses. Therefore, investigating the thermal and hygroscopic effects on a composite material is crucial.

Stresses and strains due to heat and moisture are called hygrothermal stresses and strains. The polymer matrix swells in response to changes in temperature and moisture content.

Mechanical and hydrothermal strains together make up the total strains. (2.35)

It is typical to prepare composite materials at high temperatures before cooling them to room temperature. This temperature differential is between 200°C and 300°C for composites with polymeric matrix materials and up to 1,000°C for composites with ceramic matrix materials. When it cools down, residual tensions cause a lamina because the matrix's and fiber's coefficients of thermal expansion are out of whack. The cooling process also causes the lamina to experience expansional strains. Most polymeric matrix composites can also absorb or lose moisture. Similar to those caused by thermal expansions, this moisture shift causes swelling strains and stresses. Due to the varying hygrothermal expansion of each lamina in laminates where the laminae are arranged at different angles, each lamina has residual strains. Due to differences in the elastic constants, thermal expansion coefficients, and moisture expansion coefficients of the fiber and matrix, the hygrothermal stresses in a lamina in the longitudinal and transverse directions are not equal. For unidirectional and angle laminae subject to hygrothermal loads, stress-strain relationships are developed in the sections that follow.

The strains caused by heat are given by:

$$\begin{bmatrix} \varepsilon_1^T \\ \varepsilon_2^T \\ 0 \end{bmatrix} = \Delta T \begin{bmatrix} \alpha_1 \\ \alpha_2 \\ 0 \end{bmatrix} \tag{2.36}$$

where $\begin{bmatrix} \varepsilon_1^T \\ \varepsilon_2^T \\ 0 \end{bmatrix}$ is the thermal induced strain, ΔT is the temperature change, and

$\begin{bmatrix} \alpha_1 \\ \alpha_2 \\ 0 \end{bmatrix}$ is the coefficient of thermal expansion. The moisture-induced strains

are given by:

$$\begin{bmatrix} \varepsilon_1^H \\ \varepsilon_2^H \\ 0 \end{bmatrix} = \Delta C \begin{bmatrix} \beta_1 \\ \beta_2 \\ 0 \end{bmatrix} \tag{2.37}$$

where $\begin{bmatrix} \varepsilon_1^H \\ \varepsilon_2^H \\ 0 \end{bmatrix}$ is the hygroscopic-induced strain, ΔC is the weight of moisture

absorption per unit weight of lamina, and $\begin{bmatrix} \beta_1 \\ \beta_2 \\ 0 \end{bmatrix}$ is the coefficient of hygroscopic expansion. B_1 and B_2 are the longitudinal and transverse coefficients of moisture, respectively, and ΔC is the weight of absorption per unit weight of the lamina.

$$
\begin{bmatrix} \sigma_1 \\ \sigma_{22} \\ \tau_{12} \end{bmatrix} = \begin{bmatrix} Q_{11} & Q_{12} & 0 \\ Q_{12} & Q_{22} & 0 \\ 0 & 0 & Q_{66} \end{bmatrix} \begin{bmatrix} \varepsilon_1 - \alpha_1 \Delta T - \beta_1 \Delta C \\ \varepsilon_2 - \alpha_2 \Delta T - \beta_1 \Delta C \\ \gamma_{12} \end{bmatrix} \tag{2.38}
$$

where $\begin{bmatrix} Q_{11} & Q_{12} & 0 \\ Q_{12} & Q_{22} & 0 \\ 0 & 0 & Q_{66} \end{bmatrix}$ is the reduced stiffness matrix.

$$
\begin{bmatrix} \sigma_x \\ \sigma_y \\ \tau_{xy} \end{bmatrix} = \begin{bmatrix} \bar{Q}_{11} & \bar{Q}_{12} & \bar{Q}_{16} \\ \bar{Q}_{12} & \bar{Q}_{22} & \bar{Q}_{26} \\ \bar{Q}_{16} & \bar{Q}_{26} & \bar{Q}_{66} \end{bmatrix}_k \begin{bmatrix} \varepsilon_x - \alpha_x \Delta T - \beta_x \Delta C \\ \varepsilon_y - \alpha_y \Delta T - \beta_y \Delta C \\ \gamma_x - \alpha_{xy} \Delta T - \beta_{xy} \Delta C \end{bmatrix}_k \tag{2.39}
$$

where $\begin{bmatrix} \varepsilon_x - \alpha_x \Delta T - \beta_x \Delta C \\ \varepsilon_y - \alpha_y \Delta T - \beta_y \Delta C \\ \gamma_x - \alpha_{xy} \Delta T - \beta_{xy} \Delta C \end{bmatrix}_k$ denotes the stresses in x-y co-ordinates for the

Kth layer and $\begin{bmatrix} \bar{Q}_{11} & \bar{Q}_{12} & \bar{Q}_{16} \\ \bar{Q}_{12} & \bar{Q}_{22} & \bar{Q}_{26} \\ \bar{Q}_{16} & \bar{Q}_{26} & \bar{Q}_{66} \end{bmatrix}_k$ denotes the transformed reduced stiffness

matrix. Also, stresses can be expressed as:

$$
\begin{bmatrix} \alpha_x \\ \alpha_y \\ \alpha_{xy} \end{bmatrix} = [T]^{-1} \begin{bmatrix} \alpha_1 \\ \alpha_2 \\ 0 \end{bmatrix} \tag{2.40}
$$

$$
\begin{bmatrix} \beta_x \\ \beta_y \\ \beta_{xy} \end{bmatrix} = [T]^{-1} \begin{bmatrix} \beta_1 \\ \beta_2 \\ 0 \end{bmatrix} \tag{2.41}
$$

The coefficient of thermal expansion for a unidirectional lamina is given by the transformed reduced stiffness matrix.

Similarly, B_x, B_y, and B_{xy} are the coefficients of moisture expansion for an angle lamina and are given in terms of the coefficients of moisture expansion for a unidirectional lamina.

$$[Q] = \begin{bmatrix} 20 & 0.7 & 0 \\ 0.7 & 2 & 0 \\ 0 & 0 & 0.7 \end{bmatrix} GPa$$

$\alpha_1 = 7 \times 10^{-6} /°C$

$\alpha_2 = 23 \times 10^{-6} /°C$

(a) (b)

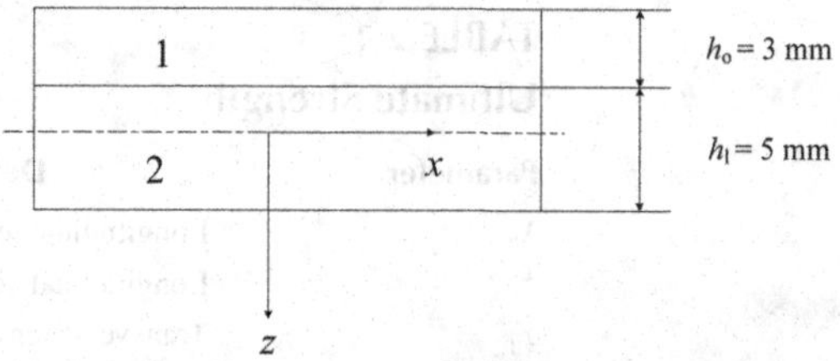

FIGURE 2.8 Two-ply laminate; (a) ABD matrix, (b) The geometry of cross section.

- **Home Work #2:**
  A two-ply laminate with ply orientations of 0° and 45° and laminate axes similar to those in Figure 2.8 should be taken into consideration. The top lamina, at 45°, is 3 mm thick, while the bottom lamina has a 0° layer and a thickness of 5 mm. Calculate the resulting moments (NT, MT) and forces (NT, MT) in the laminate that is manufactured at 125°C and cooled to 25°C. Both laminae have the same stiffness matrix Q, as shown in Figure 2.8:

2.4 PIEZOELECTRIC COMPOSITES

Piezocomposites, which combine a piezoelectric ceramic with a polymer, are fascinating materials because of their high degree of tailorability. Based on the connection of each phase, ten variants of the geometry of two-phase composites are possible (one-, two-, or three-dimensionally). The benefits of a 1–3 piezocomposite (PZT-rod/polymer-matrix composite) are enormous. Advantages include a wide bandwidth, low mechanical quality factor, low acoustic impedance, high coupling factors, good matching to water or human tissue, and mechanical flexibility. Applications for piezoelectric composite materials in ultrasonic transducers include underwater sonar and medical diagnostic ultrasound.

Piezoelectric passive dampers are more successful in reducing noise and vibration than conventional rubbers because they are built of a piezoelectric ceramic particle, a polymer, and carbon black. A combination of a magnetostrictive ceramic and a piezoelectric ceramic produces the magnetoelectric effect, in which an electric field is generated in the material in response to an applied magnetic field.

2.5 FAILURE ANALYSIS IN COMPOSITE MATERIALS

Failure mechanism in composite materials is more complicated than in isotropic materials. There are several types of failure in a composite structure, namely, fiber buckling, fiber breakage, matrix cracking, and delamination. Most failure criteria for composite materials cannot predict the type of failure. These criteria

TABLE 2.1
Ultimate Strength

Parameter	Definition
X_T	Longitudinal tensile strength
X_C	Longitudinal compressive strength
Y_T	Transverse tensile strength
Y_C	Transverse compressive strength
S	Shear strength

are based on the failure in each single orthotropic lamina of a laminated composite. Therefore, we need to employ the criteria for each single layer and find the first ply that will fail. To obtain the failure occurrence probability in a lamina, all applied stresses should be calculated in principal material coordinate.

2.5.1 MAXIMUM STRESS FAILURE CRITERION

In the maximum stress failure criterion, each and every one of the stresses in principal material coordinates must be less than the respective strength. For tensile stresses $\sigma_1 < X_t$ $\sigma_2 < Y_t$, for compression stresses $\sigma_1 > X_c$ $\sigma_2 > Y_c$, and for shear stresses $|\tau_{12}| < S$. The definitions of parameters are provided in Table 2.1.

2.5.2 MAXIMUM STRAIN FAILURE CRITERION

The maximum strain failure criterion and the previous criterion are pretty similar. But in this case, strains rather than stresses are constrained. Any one or more of the following inequalities must be satisfied for the material to pass.

For tensile strains $\varepsilon_1 < X_{\varepsilon t}$ $\varepsilon_2 < Y_{\varepsilon t}$, for compression strains $\varepsilon_1 > X_{\varepsilon c}$ $\varepsilon_2 > Y_{\varepsilon c}$, and for shear strains $|\gamma_{12}| < S_\varepsilon$. $X_{\varepsilon t}(X_{\varepsilon c})$ is the maximum tensile (compressive) normal strain in the direction 1. $Y_{\varepsilon t}(Y_{\varepsilon c})$ is the maximum tensile (compressive) normal strain in the direction 2.

S_ε is the maximum shear strain in the 1–2 coordinates.

2.5.3 CONVENTIONAL FAILURE THEORIES FOR COMPOSITE MATERIALS

1. **Tsai-Hill failure theory:** The Tsai-Hill failure theory considers the interaction between the lamina strength parameters unlike the maximum strain and stress failure theories. The appropriate values of X_T, X_C, Y_T, and Y_C must be used instead of X and Y depending on the sign of σ_1 and σ_2:

$$\frac{\sigma_1^2}{X^2} - \frac{\sigma_1 \sigma_2}{X^2} + \frac{\sigma_2^2}{Y^2} + \frac{\tau_{12}^2}{S^2} < 1 \tag{2.42}$$

2. **Hoffman failure theory:** Instead of individual criteria for failure like the maximum stress criterion, interactions between failure modes are taken into consideration. The Hoffman criterion is the most straightforward criterion used in the design:

$$\frac{\sigma_1^2}{X_c X_t} + \frac{\sigma_1 \sigma_2}{X_c X_t} - \frac{\sigma_2^2}{Y_c Y_t} + \frac{X_c + X_t}{X_c X_t}\sigma_1 + \frac{Y_c + Y_t}{Y_c Y_t}\sigma_2 + \frac{\tau_{12}^2}{S_{12}^2} < 1 \qquad (2.43)$$

where definitions of parameters are provided in Table 2.1.

- **Home Work #3:**

    Consider a symmetric cross-ply square composite plate as shown in Figure 2.9 with (0°/90°/0°) stacking sequence. The layers have identical thickness, 1 mm, and elastic and strength properties as presented in Tables 2.2 and 2.3. The plate size is 0.25 m.

Determine the failure probability for the plate and the first failed ply for each mentioned criterion. Compare the results in a table.

Repeat the problem for a symmetric angle-ply laminate and $\alpha = 45°$.

What are the effects of changing layup on the plate's failure?

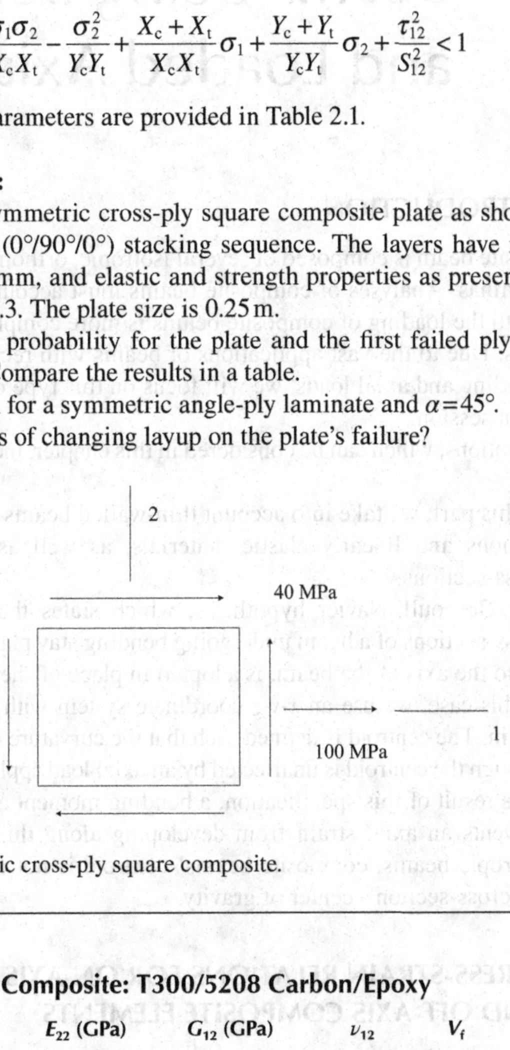

FIGURE 2.9 Symmetric cross-ply square composite.

TABLE 2.2

Elastic Properties of Composite: T300/5208 Carbon/Epoxy

Material	E_{11} (GPa)	E_{22} (GPa)	G_{12} (GPa)	ν_{12}	V_f	ρ (kg/m³)
UD	162.0	14.9	5.7	0.283	0.7	1,583

TABLE 2.3

Strength Properties of Composite: T300/5208 Carbon/Epoxy

Material	X_T (MPa)	Y_T (MPa)	S_{12} (MPa)	X_c (Mpa)	Y_c (MPa)
UD	1,744	52.6	107.8	1,650	260

3 Rectangular Composite Beams Being Bended and Loaded Axially

3.1 INTRODUCTION

A composite beam is composed of several isotropic, orthotropic, and woven composite laminas. Analyses of composite beams must account for the fact that the response to the loading of composite beams is more complex than that of isotropic beams. Due to the vast applications of beams with rectangular cross-section under bending and axial loads, we will focus on this type of composite beams in the current session.

Assumptions, which can be considered in this chapter, include the following [1]:

1. In this part, we take into account thin-walled beams with modest deformations and linearly elastic materials, as well as rectangular solid cross-sections.
2. The Bernoulli-Navier hypothesis, which states that the initial plane cross-sections of a beam undergoing bending stay plane and perpendicular to the axis of the beam, is adopted in place of shear deformations.
3. In this case, we use an x-y-z coordinate system with the centroid as the origin. The centroid is defined such that the curvature of the axis traveling through the centroid is unaffected by an axial load applying at the centroid.
4. As a result of this specification, a bending moment applied to the beam prevents an axial strain from developing along this axis. Contrary to isotropic beams, composite beams' centroid is not always the same as the cross-section's center of gravity.

3.2 STRESS-STRAIN RELATIONS FOR ON-AXIS AND OFF-AXIS COMPOSITE ELEMENTS

Think about a component of a unidirectional on-axis composite or a composite whose primary material axes (1, 2, and 3) are in line with the coordinate system. The stresses are volume averages over the fiber and matrix domains, as shown in Figures 3.1–3.3, which also displays the definition of the stress components associated with the material coordinate system. The normal stresses are σ_1, σ_2, and σ_3, while the shear stresses are τ_{12}, τ_{13}, and τ_{23}. Corresponding normal strains are ε_1, ε_2, and ε_3, and the engineering shear strains are γ_{12}, γ_{13}, and γ_{23}.

DOI: 10.1201/9781003429197-3

In thin, sheet-like structures such as a ply in a laminate, it is common to assume a state of plane stress by setting $\sigma_3 = \tau_{13} = \tau_{23} = 0$.

It may be demonstrated that such a level of stress causes the out-of-plane shear strains to dissipate, i.e., $\gamma_{13} = \gamma_{23} = 0$. The out-of-plane extensional strain, ε_3, does not vanish but becomes coupled to the in-plane stresses σ_1 and σ_2 and does not remain an independent quantity. The stress-strain relation for plane stress becomes:

$$
\begin{matrix}
\sigma_1 \\
\sigma_2 \\
\tau_{12}
\end{matrix}
\ = \
\begin{matrix}
Q_{11} & Q_{12} & 0 \\
Q_{12} & Q_{22} & 0 \\
0 & 0 & Q_{66}
\end{matrix}
\begin{matrix}
\varepsilon_1 \\
\varepsilon_2 \\
\tau_{12}
\end{matrix}
\tag{3.1}
$$

where the reduced stiffnesses, Q_{ij}, can be expressed in terms of constants as shown in Figure 3.1 and the following equations:

$$
Q_{11} = \frac{E_1}{1 - \nu_{12}\nu_{21}}
\tag{3.2}
$$

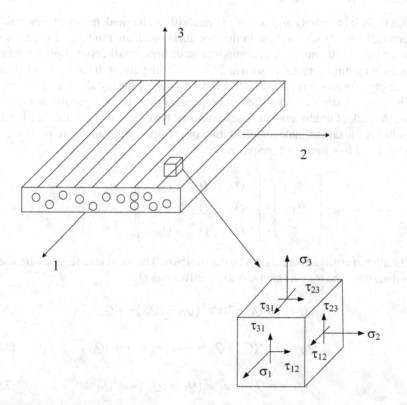

FIGURE 3.1 Components of normal and shear stress definition and on-axis composite element.

$$Q_{12} = \frac{v_{12}E_2}{1 - v_{12}v_{21}} \tag{3.3}$$

$$Q_{22} = \frac{E_2}{1 - v_{12}v_{21}} \tag{3.4}$$

$$Q_{66} = G_{12} \tag{3.5}$$

where E_1 and E_2 are the principal Young's moduli, and G_{12} is the in-plane shear modulus. v_{12} and v_{21} are the principal (major and minor) Poisson ratios. Utilizing woven fabric composites, where thousands of fibers are organized in a certain pattern, such as "one fiber bundle over-one under" in plain weave fabrics, is a typical strategy to obtain a set of more balanced mechanical qualities. Other weave patterns exist in addition to the plain weave fiber pattern. Equation (3.5) can also be used to illustrate the mechanical behavior of a plain weave composite layer made of fibers oriented in 0° and 90°.

3.2.1 OFF-AXIS SYSTEM

Due to their incredibly anisotropic characteristics and weak failure planes, unidirectional composites, as those in the aforementioned, are rarely used in practical structures. In the majority of composite structures, unidirectional plies are oriented in multiple directions, such as 0°, 45°, 90°, and others. It has been noted that each ply orientation is produced by properly rotating the ply about its three axes. The stresses, strains, and stiffness components must be changed for an off-axis ply. A stack of unidirectional composites that have been bonded together form a multidirectional laminate and definition of ply orientation angle are provided in parts a and b of figure 3.2, respectively.

$$\begin{bmatrix} \sigma_x \\ \sigma_y \\ \tau_{xy} \end{bmatrix} = \begin{bmatrix} \overline{Q}_{11} & \overline{Q}_{12} & \overline{Q}_{16} \\ \overline{Q}_{12} & \overline{Q}_{22} & \overline{Q}_{26} \\ \overline{Q}_{16} & \overline{Q}_{26} & \overline{Q}_{66} \end{bmatrix} \begin{bmatrix} \varepsilon_X \\ \varepsilon_Y \\ \Upsilon_{XY} \end{bmatrix} \tag{3.6}$$

The altered attributes are shown by the overbars. The on-axis stiffnesses are used to determine the transformed (off-axis) stiffnesses Q_{ij}:

$$\overline{Q}_{11} = m^4 Q_{11} + 2m^2 n^2 (Q_{12} + 2Q_{66}) + n^4 Q_{22} \tag{3.7}$$

$$\overline{Q}_{12} = m^2 n^2 (Q_{11} + Q_{22} - 4Q_{66}) + (m^4 + n^4) Q_{12} \tag{3.8}$$

$$\overline{Q}_{22} = n^4 Q_{11} + 2m^2 n^2 (Q_{12} + 2Q_{66}) + m^4 Q_{22} \tag{3.9}$$

$$\overline{Q}_{16} = m^3 n (Q_{11} - Q_{12}) + mn^3 (Q_{12} - Q_{22}) - 2mn(m^2 - n^2) Q_{66} \tag{3.10}$$

$$\bar{Q}_{26} = mn^3\left(Q_{11} - Q_{12}\right) + m^3n\left(Q_{12} - Q_{22}\right) + 2mn\left(m^2 - n^2\right)Q_{66} \qquad (3.11)$$

$$\bar{Q}_{66} = m^2n^2\left(Q_{11} + Q_{22} - 2Q_{12} - 2Q_{66}\right) + \left(m^4 - n^4\right)Q_{66} \qquad (3.12)$$

where $m = \cos\theta$ and $n = \sin\theta$.

The compressed form of the laminate's stiffness for a sandwich panel can be represented as:

$$\begin{bmatrix} N \\ M \end{bmatrix} = \begin{bmatrix} A & B \\ C & D \end{bmatrix}\begin{bmatrix} \varepsilon^\circ \\ k \end{bmatrix} \qquad (3.13)$$

By inverting the 6×6 $ABCD$ matrix, it is sometimes possible to represent the core midplane strains and curvatures in terms of force and moment resultants:

$$\begin{bmatrix} \varepsilon^\circ \\ k \end{bmatrix} = \begin{bmatrix} a & b \\ c & d \end{bmatrix}\begin{bmatrix} N \\ M \end{bmatrix} \qquad (3.14)$$

where $[a]$, $[b]$, $[c]$, and $[d]$ are the 3×3 compliance matrices given by:

$$[a] = \left[A^*\right] - \left[B^*\right]\left[D\right]^{-1}\left[C^*\right] \qquad (3.15)$$

$$[b] = \left[B^*\right]\left[D^*\right]^{-1} \qquad (3.16)$$

$$[c] = -\left[D^*\right]^{-1}\left[C^*\right] \qquad (3.17)$$

$$[d] = \left[D^*\right]^{-1} \qquad (3.18)$$

$$\left[A^*\right] = \left[A^*\right]^{-1} \qquad (3.19)$$

$$\left[B^*\right] = -\left[A^*\right]^{-1}\left[B\right] \qquad (3.20)$$

$$\left[C^*\right] = [C][A]^{-1} \qquad (3.21)$$

$$\left[D^*\right] = [D] - [C]\left[A^*\right]^{-1}[B] \qquad (3.22)$$

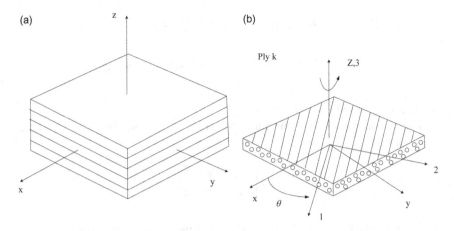

FIGURE 3.2 (a) A stack of unidirectional composites that have been bonded together form a multidirectional laminate and (b) definition of ply orientation angle.

It was demonstrated that $B_{ij} = C_{ij}$, which results in the particular scenario examined in Section 3.2, when the faces are viewed as homogenous orthotropic materials:

$$\left[D^*\right] = [D] - [C]\left[A^*\right]^{-1}[B] \tag{3.23}$$

$$\left[A^*\right] = [A]^{-1} \tag{3.24}$$

$$\left[B^*\right] = -[A]^{-1}[B] \tag{3.25}$$

$$\left[C^*\right] = [B][A]^{-1} \tag{3.26}$$

$$\left[D^*\right] = [D] - [B][A]^{-1}[B] \tag{3.27}$$

In addition, $[B] = [C] = [0]$ for the exceptional situation of a symmetric sandwich (Section 3.2), i.e., where the face sheets are identical and laid up with their midplane as a mirror plane, is true. In turn, $[B] = [C] = [0]$ is the outcome:

$$[a] = [A]^{-1} \tag{3.28}$$

$$[b] = [c] = [0] \tag{3.29}$$

$$[d] = [D]^{-1} \tag{3.30}$$

Furthermore, it may be demonstrated that, for a sandwich with no D_{16} and D_{26} terms:

$$d_{11} = \frac{D_{22}}{D_{11}D_{22} - D_{12}^2} \tag{3.31}$$

$$d_{12} = \frac{-D_{12}}{D_{11}D_{22} - D_{11}^2} \tag{3.32}$$

$$d_{22} = \frac{D_{11}}{D_{11}D_{22} - D_{12}^2} \tag{3.33}$$

$$d_{66} = \frac{1}{D_{66}} \tag{3.34}$$

3.3 HOOKE'S LAW

For a material, the generalized Hooke's law is given as:

$$\sigma_{IJ} = C_{ijkl}\varepsilon_{kl} \quad i,j,k,l = 1,2,3 \tag{3.35}$$

The stress components make up the second-order tensor σ_{ij}, sometimes known as the stress tensor. The second-order tensor ij is also referred to as the strain tensor, and the strain components make up each of its constituent parts. The stiffness tensor, or C_{ijkl}, is a fourth-order tensor. As it is commonly known, we will refer to it as a stiffness matrix in the following section. The stiffness coefficients for this linear stress-strain relationship are represented by each of the tensor's component elements. As a result, each of the stress and strain tensors has three components $(3\times3=9)$, while the stiffness tensor has 81 independent parts. Elastic constants, such as moduli and stiffness coefficients, are different names for the separate components. With the help of the following symmetries, the number of these elastic constants can be decreased.

Stress symmetry: The stress components are symmetric under this symmetry condition, that is, $\sigma_{ij}=\sigma_{ji}$. Thus, there are six independent stress components:

$$\sigma_{ji} = C_{jikl}\varepsilon_{kl} \tag{3.36}$$

Subtracting Eq. (3.22) from Eq. (3.21):

$$0 = \left(C_{ijkl} - C_{jikl}\right)\varepsilon_{kl} \rightarrow C_{ijkl} = C_{ijlk} \tag{3.37}$$

i and j can be expressed in six different ways on their own, and K and L can be expressed in nine different ways on their own. As a result, the symmetry reduces the 81 independent elastic constants to 54.

3.3.1 STRAIN SYMMETRY

The strain components are symmetric under this symmetry condition, that is, ij ji $\varepsilon = \varepsilon$. Hence, from Eq. (3.21) it can be written as:

$$\sigma_{ij} = C_{ijlk}\varepsilon_{lk} \tag{3.38}$$

Subtracting Eq. (3.23) from Eq. (3.22), we get the following equation:

$$0 = \left(C_{ijkl} - C_{ijlk}\right)\varepsilon_{kl} \rightarrow C_{ijkl} = C_{ijlk} \tag{3.39}$$

Equation (3.3) demonstrates that while k and l are fixed, there are six different methods to describe I and j taken together. Similar to this, when I and j in Eq (3.39). are fixed, there are six different methods to describe k and l taken together (Eq. 3.4). Because of the symmetry of the stress and strain, this linear elastic material has $6\times6=36$ independent constants. With fewer stiffness coefficients and decreased stress and strain components, Hooke's rule can be expressed in a condensed form as:

$$\sigma_i = C_{ij}\varepsilon_j \quad (i, j = 1, 2, \ldots, 6) \tag{3.40}$$

where,

$$
\begin{aligned}
\sigma_1 &= \sigma_{11} & \varepsilon_1 &= \varepsilon_{11} \\
\sigma_2 &= \sigma_{22} & \varepsilon_2 &= \varepsilon_{22} \\
\sigma_3 &= \sigma_{33} & \varepsilon_3 &= \varepsilon_{33} \\
\sigma_4 &= \sigma_{23} & \varepsilon_4 &= 2\varepsilon_{23} \\
\sigma_5 &= \sigma_{13} & \varepsilon_5 &= 2\varepsilon_{13} \\
\sigma_6 &= \sigma_{12} & \varepsilon_6 &= 2\varepsilon_{12}
\end{aligned} \tag{3.41}
$$

It should be highlighted that the strains in this case are engineering strains. The determinant of the stiffness matrix must be nonzero for Eq. (3.25) to be solvable for strains in terms of stresses, that is, $|C_{ij}| \neq 0$.

If there is a strain energy density function W, as shown below, the number of distinct elastic constants can be further decreased.

Given as follows is the strain energy density function:

$$W = \frac{1}{2}C_{ij}\varepsilon_i\varepsilon_j \tag{3.42}$$

$$\sigma_i = \frac{\delta W}{\delta\varepsilon_i} \tag{3.43}$$

We can observe that the function of strain on W is quadratic. A substance that possesses W and the quality in Eq. (3.26) is referred to as a hyperelastic substance. Another way to write the W is as:

$$W = \frac{1}{2} C_{ji} \varepsilon_j \varepsilon_i \qquad (3.44)$$

Subtracting Eq. (3.27) from Eq. (3.26), we get:

$$0 = \left(C_{ij} - C_{ji}\right) \varepsilon_i \varepsilon_j \qquad (3.45)$$

It results in $C_{ij} = C_{ji}$ as an identity. The stiffness matrix is hence symmetric. There are 21 independent elastic constants in this symmetric matrix. The following is the stiffness matrix:

$$C_{ij} = \begin{bmatrix} C_{11} & C_{12} & C_{13} & C_{14} & C_{15} & C_{16} \\ & C_{22} & C_{23} & C_{24} & C_{25} & C_{26} \\ & & C_{33} & C_{34} & C_{35} & C_{36} \\ & & & C_{44} & C_{45} & C_{46} \\ & \text{Symmetric} & & & C_{55} & C_{56} \\ & & & & & C_{66} \end{bmatrix} \qquad (3.46)$$

The first and second laws of thermodynamics are the foundation for the function W's existence. The fact that this function is positive definite should also be highlighted. The function W is an invariant as well. (An invariant is a quantity that is independent of the change of reference.) A material is referred to as anisotropic or aelotropic if it has 21 independent elastic constants.

3.4   USING THE STRENGTH OF MATERIALS APPROACH, BEND CURVED BEAMS

We have been researching initially straight members up until this point. In this chapter, we'll look at how originally curved beams can bend. To achieve this, we limit our analysis to situations in which bending occurs in the plane of curvature. This occurs when the bending force acts in the plane of the beam's curvature and the cross-section of the beam is symmetrical about that plane. We first arrive at the answer, as we did for straight beams, assuming that originally planar parts stay that way after bending [2]. Hooke's law is the name given to the relationship that results between stress, deflection, and deformation. When a curved beam is exposed to a pure bending moment or end load, we can still derive the stress and displacement field using the two-dimensional elasticity formulation without assuming that plane parts remain plane. In order to come to a conclusion, we compare the two answers and discover that, when the beam is shallow, they are in perfect accord.

Be clear on what is meant by a curved beam before continuing. A beam is said to be curved if its axis is not straight and curves upward. The beam's axis should be curved in the x-y plane if the applied loads are distributed along the y direction and the beam's span is distributed along the x direction. On the contrary, if the member is bent in the x-y plane while still being loaded along the y axis, this is not a curved beam because the loading will cause the section to bend and twist. Consequently, a curved beam lacks a curvature in the plan. Curved beams include structures like arches as shown in Figure 3.3.

Assume that straight beams have straight portions that only rotate along the neutral axis and that the hoop stress is the only meaningful stress $\sigma_{\theta\theta}$ as shown in Figure 3.4.

From Hook's law:

$$\sigma_{\theta\theta} = Ee_{\theta\theta} = E_w\left(\frac{R_n}{r} - 1\right) \tag{3.47}$$

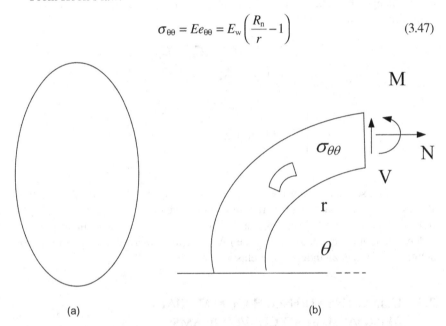

(a) (b)

FIGURE 3.3 (a) Symmetric cross-section and (b) curved beam with applied moment M.

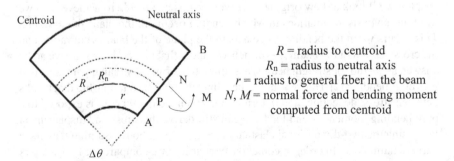

R = radius to centroid
R_n = radius to neutral axis
r = radius to general fiber in the beam
N, M = normal force and bending moment computed from centroid

FIGURE 3.4 Centroid and neutral axis for a curved beam.

Then, the normal force is given by:

$$N = \int_A \sigma_{\theta\theta} dA = E_w \left[R_n \int_A \frac{dA}{r} - \int_A dA \right] = E_w \left(R_n A_m - A \right) \quad (3.48)$$

where $A_n = \int_A \frac{dA}{r}$ has the dimensions of a length.

$$M = \int_A \sigma_{\theta\theta} (R - r) dA$$

$$= En \int_A \left(\frac{R_n}{r} - 1 \right) (R - r) dA \quad (3.49)$$

$$= En \left[R_n R \int_A \frac{dA}{r} - R \int_A dA - R_n \int_A dA + \int_A r dA \right]$$

$$= En R_n (R A_m - A)$$

$$M = E_w R_n (R A_m - A) \quad (3.50)$$

$$N = E_w \left(R_n A_m - A \right) \quad (3.51)$$

From (Eq. 3.32):

$$E_w R_n = \frac{M}{R A_m - A} \quad (3.52)$$

From (Eq. 3.33):

$$N = (E_w R_n) A_m - E_w A = \frac{M A_m}{R A_m - A} - E_w A \quad (3.53)$$

So, solving for E_w:

$$E_w = \frac{M A_m}{A (R A_m - A)} - \frac{N}{A} \quad (3.54)$$

Recall that the stress is given by:

$$\sigma_{\theta\theta} = E e_{\theta\theta} = E_w \left(\frac{R_n}{r} - 1 \right) = \frac{E_w R_n}{r} - E_w \quad (3.55)$$

Applying expressions for $E_w R_n, Ew$, we calculate the hoop stress:

$$\sigma_{\theta\theta} = \frac{N}{A} + \frac{M(A - rA_m)}{Ar(RA_m - A)} \tag{3.56}$$

Bending stress and axial stress:

$$N \neq 0 \tag{3.57}$$

Setting the total stress$=0$ gives:

$$r\,|_\sigma = \frac{AM}{A_m M + N(A - RA_m)} \tag{3.58}$$

$$N = 0$$

Changing the bending stress to 0 and $r = R_n$ gives $R_n = \dfrac{A}{A_m}$ location of the neutral axis, which in general is not at the centroid.

Example 3.1

For a square of 50 × 50 mm area, calculate the maximum tensile and compressive stress if $P=9.5$ kN and graph the total stress across the cross-section as shown in Figure 3.5.

Solution

$$A = (50)(50) = 2,500\,\text{mm}^2$$

so we have $\quad A_m = 50 \ln\left(\dfrac{80}{30}\right) = 49.04\,\text{mm} \qquad R_n = \dfrac{2,500}{49.04} = 51\,\text{mm}$

$$R = \frac{80 + 30}{2} 55\,\text{mm}$$

Max tensile stress is at $r=30$ mm.

$$\sigma_{\theta\theta} = \frac{N}{A} + \frac{M(A - rA_m)}{Ar(RA_m - A)}$$

$$= \frac{9,500}{2,500} + \frac{(155)(9,500)\left[2,500 - (30)(49.04)\right]}{(2,500)(30)\left[(55)(49.04) - 2,500\right]}$$

$$= 106.2\,\text{MPa}$$

Max compressive stress is at 80 mm.

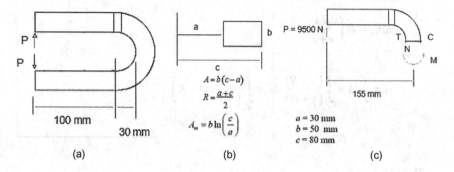

FIGURE 3.5   Cross-section of Example 3.1; (a) Part A, (b) Part B, and (c) Part C.

$$\sigma_{\theta\theta} = \frac{N}{A} + \frac{M(A - rA_m)}{Ar(RA_m - A)}$$

$$= \frac{9,500}{2,500} + \frac{(155)(9,500)[2,500 - (80)(49.04)]}{(2,500)(80)[(55)(49.04) - 2,500]}$$

$$= -49.3\,\text{MPa}$$

On Juxtaposing with Bickford's expression for pure bending:

$$\sigma_{\theta\theta} = -\frac{kM}{A} - \frac{My}{(1 + ky)I_2}$$

$k = 1/R$, here, distance from the centroid is y.

$$\int_A y\,dA = 0$$

So

$$\int_A \frac{y\left(1 + \dfrac{y}{R}\right)}{1 + \dfrac{y}{R}}\,dA = \int_A \frac{y\,dA}{1 + \dfrac{y}{R}} + \frac{1}{R}\int_A \frac{y^2\,dA}{1 + \dfrac{y}{R}} = I_1 + \frac{I_2}{R} = 0$$

$$\rightarrow I_1 = -\frac{I_2}{R}$$

$$I_1 = \int_A \frac{y\,dA}{1 + \dfrac{y}{R}}$$

$$I_2 = \int_A \frac{y^2\,dA}{1 + \dfrac{y}{R}}$$

$$RA_\mathrm{m} = R \int_A \frac{dA}{y+R}$$

$$A = \int_A dA = \int_A \frac{(y+R)}{(y+R)} dA = \int_A \frac{y}{y+R} dA + R \int_A \frac{dA}{y+R}$$

$$= \frac{I_1}{R} + RA_\mathrm{m}$$

Thus:

$$R(RA_\mathrm{m} - A) = -I_1 = \frac{I_2}{R}$$

Now, start with Bickford's expression:

$$\sigma_{\theta\theta} = -\frac{kM}{A} - \frac{My}{(1+ky)I_2}$$

$$k = \frac{1}{R}$$

$$\sigma_{\theta\theta} = -M\left[\frac{1}{AR} + \frac{yR}{(y+R)I_2} \right]$$

$$= -M\left[\frac{(y+R)I_2 + yAR^2}{(y+R)ARI_2} \right]$$

$$= -M\left[\frac{(y+R)I_2 + (y+R)AR^2}{(y+R)ARI_2} - \frac{AR^3}{(y+R)ARI_2} \right]$$

$$= -M\left[\frac{I_2 + AR^2}{ARI_2} - \frac{R^2}{(y+R)I_2} \right]$$

$$= M\left[-\frac{I_2 + AR^2}{ARI_2} + \frac{R^2}{(y+R)I_2} \right]$$

$$= M\left[\frac{A - \dfrac{(y+R)(I_2 + AR^2)}{R^3}}{\dfrac{(y+R)AI_2}{R^2}} \right]$$

$$RA_\mathrm{m} - A = \frac{I_2}{R^2},$$

Therefore:

$$A_m = \frac{I_2 + AR^2}{R^3}$$

We get that:

$$\sigma_{\theta\theta} = \left[\frac{A - rA_m}{r(RA_m - A)A} \right]$$

From Bickford's expression

$$\sigma_{\theta\theta} = -\frac{kM}{A} - \frac{My}{(1+ky)I_2}$$

$$k - \frac{1}{R}$$

We can see that $R \to \infty$, $k \to 0$

And

$$I_2 = \int_A \frac{y^2 dA}{1 + \frac{y}{R}} \to \int_A y^2 dA = I$$

Here, we obtain the straight beam flexure equation:

$$\sigma_{\theta\theta} = -\frac{My}{I}$$

The radial and tangential stresses were calculated by the following equations:

$$\sigma_r = \frac{4M}{tb^2 N} \left[\left(1 - \frac{a^2}{b^2}\right) \cdot \left(\ln \frac{r}{a}\right) - \left(1 - \frac{a^2}{r^2}\right) \cdot \left(\ln \frac{b}{a}\right) \right]$$

$$\sigma_r = \frac{4M}{tb^2 N} \left[\left(1 - \frac{a^2}{b^2}\right) \cdot \left(1 + \ln \frac{r}{a}\right) - \left(1 + \frac{a^2}{r^2}\right) \cdot \left(\ln \frac{b}{a}\right) \right]$$

Where, $N = \left(1 - \frac{a^2}{b^2}\right) - 4\frac{a^2}{b^2} \ln^2 \left(\frac{b}{a}\right)$

3.5 FEM SIMULATION MODELS

This chapter explains and examines how a number of virtual models were created to help people understand the mechanics of materials. The simulated models are built using the finite element program Abaqus. When there is no physical lab to support the study of a material's mechanics, the models can be particularly useful. In order to more clearly demonstrate the value and significance of the models, this chapter presents a few examples of the simulations that were created. Simulations can be used, for example, to calculate and display the stress and deformation contours at different points on solid continuums with changing geometries, boundary conditions, material properties, and loading circumstances. Examples given in this chapter include the analysis of axially loaded members and beam bending.

3.5.1 AXIAL LOADING

A-36 steel plate with a round hole and a fillet, which is thin and 5 mm thick, is supported on the left by a permanent support and is put under a 20 MPa uniformly distributed tensile load at the right end. The plate has an elastic modulus (E) of 200 GPa and a Poisson's ratio (n) of 0.32. To assist in meshing the area, ten partitions were created on the model in Abaqus. The partitions assist in generating a finer mesh where stress concentrations are present, such as in the vicinity of the fillet and the hole.

The analysis uses an eight-node biquadratic plane stress quadrilateral, decreased integration element, to get the results that are shown. The users of Abaqus can generate and visualize the stress distribution along any specified path using a variety of tools. The maximum normal stress at this section in accordance with Eq. can be obtained by using this plot to calculate the stress concentration factor K (3.34). The K value that was determined by this analysis can be compared to the stress concentration factor reported in many mechanics texts as a theoretical number [3]:

$$\sigma_{\max} = K\sigma_{\text{avg}} \tag{3.59}$$

Similar distribution can be shown in close proximity to the fillet by displaying a vertical section perpendicular to the loading axis. A comparison between the analytically determined value of K and the theoretical values is also conceivable. Many mechanics texts discuss the distributions of the parameter K that correlate to various r/h and w/h ratios. A finer mesh and more partitions could be employed to create results that are more accurate. The developed model can also be used to demonstrate that the normal stress (and normal strain) distribution is largely homogeneous away from both the fixed support and the location of the hole and fillets. To learn more about the real behavior of axially loaded structural components, Abaqus' visualization module's unique tool can be used to probe values of stress or displacement at different points on the part.

3.5.2 BENDING

In order to analyze the behavior of this structural member under such loading, this chapter also created a model of an initially curved beam that had been subjected to a bending moment. The normal stress results acting on a cross-section of the model can be compared to the theoretical values using the theoretical equation (Eq. 3.35):

$$\sigma = \frac{M(R-r)}{Ar(\bar{r}-R)}, \quad \text{where} \quad R = \frac{A}{\int_A \dfrac{dA}{r}} \tag{3.60}$$

where,

σ = Normal stress affecting the cross-section of the beam

M = Internal moment about the cross-neutral section's axis (positive if it increases the radius of curvature of the beam)

R = Distance from the neutral axis to the center of curvature

$\bar{r}$ = The distance between the cross-section's centroid and the curve's center

r = Distance between the curvature's center and the location where the normal stress must be calculated

A = Area of the beam

In this example, the model was meshed using a C3D20R type element (a 20-node quadratic brick, reduced integration element).

The beam cross-section was sliced so that the stress values acting on it along the x-direction perpendicular to the y-z plane could be seen clearly. The results produced from these stresses can be compared to those derived using the theoretical equation previously supplied (Eq. 3.35). The stress result for one sample probed element on the cross-section, which was produced using a unique tool in the visualization module of Abaqus. For other curved members with various other cross-sections, comparable models can be created. Numerous mechanics textbooks have the formulae for the bending stress of curved beams with triangular, circular, and elliptical cross-sections [4,5].

REFERENCES

[1] A. Elmarakbi, *Advanced Composite Materials for Automotive Applications: Structural Integrity and Crashworthiness*, First Edition. John Wiley & Sons, Ltd. DOI:10.1002/9781118535288, 1992, New York.

[2] A. P. Boresi, R. J. Schmidt, and O. M. Sidebottom, *Advanced Mechanics of Materials*. John Wiley & Sons.

[3] R. C. Hibbeler, *Mechanics of Materials*, Ninth Edition, Pearson Prentice Hall, 2014.

[4] S. Timoshenko, and J.N. Goodier, *Theory of Elasticity*, Third Edition, McGraw-Hill, 1934, New York.

[5] R. C. Hibbeler, *Structural Analysis*, Ninth Edition, Pearson Prentice Hall, 2012.

4 Composite Beams

4.1 CONCEPTUAL DESIGN

Shear and bending are the two halves of a composite beam's deflection ($\delta = \delta_b + \delta_s$). The bending stiffness (EI) governs the bending deflection (δ_b), whereas the shear stiffness governs the shear deflection (δ_s) (GA). Metals have a high shear modulus ($GE = E/2.5$), making shear deformations unnecessary for metallic beams, but composite materials have a low shear modulus ($E/10$ or less). The span affects how significant the shear deflection is in relation to the bending deflection; the bigger the span, the less significant the shear deflection (compared to bending). At the location where the dummy load is placed, the deflection is calculated as:

$$\delta = \int_0^L \left(\frac{M}{EI} M^0 + \frac{V}{GA} V^0 \right) dx \qquad (4.1)$$

4.2 DEFLECTION DESIGN

In the early stages of design, the carpet plot may be used to determine the average elastic modulus. The section's geometry may then be tailored to obtain the necessary bending stiffness. The laminate modulus E_x of various laminates may be calculated using carpet plots as a function of the number of 0°, 90°, and ($\pm\theta$) laminae in the laminate. The maximum modulus E_x is obtained by choosing laminates with high values, but such laminates may have low shear moduli G_{xy}, leading to undesirable levels of shear deflection. A laminate with a high value will also have a low shear strength F_{xy}, which could not be sufficient to support the shear stresses. Typically, the geometry is scaled first, and then, the shear deflection is calculated. If the shear deflection is too great, the geometry or lamination must be modified. The section's shear stiffness (GA) regulates the shear deflection. For the sake of preliminary design, it is presumed that the flanges have no impact on shear stiffness. Inferring from the carpet plot, G_{xy} represents the apparent shear modulus of the laminate and A represents the area of the webs, respectively. Greater values of G_{xy} result from choosing a laminate with higher values (in the carpet plots). If such configurations can be produced, other laminates can be used to create the webs and flanges. For instance, because of manufacturing restrictions, a rectangular tube formed by filament winding must have the same laminate structure on all its walls. The designer will work to maximize Exon in the flanges and G_{xy} in the webs if alternative laminates may be employed (Figure 4.1).

DOI: 10.1201/9781003429197-4

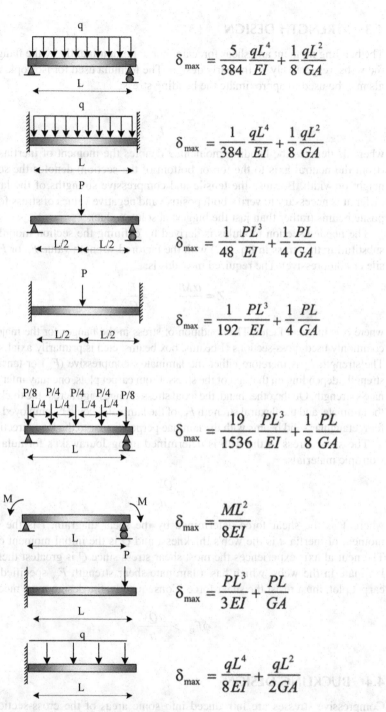

$$\delta_{max} = \frac{5}{384}\frac{qL^4}{EI} + \frac{1}{8}\frac{qL^2}{GA}$$

$$\delta_{max} = \frac{1}{384}\frac{qL^4}{EI} + \frac{1}{8}\frac{qL^2}{GA}$$

$$\delta_{max} = \frac{1}{48}\frac{PL^3}{EI} + \frac{1}{4}\frac{PL}{GA}$$

$$\delta_{max} = \frac{1}{192}\frac{PL^3}{EI} + \frac{1}{4}\frac{PL}{GA}$$

$$\delta_{max} = \frac{19}{1536}\frac{PL^3}{EI} + \frac{1}{8}\frac{PL}{GA}$$

$$\delta_{max} = \frac{ML^2}{8EI}$$

$$\delta_{max} = \frac{PL^3}{3EI} + \frac{PL}{GA}$$

$$\delta_{max} = \frac{qL^4}{8EI} + \frac{qL^2}{2GA}$$

FIGURE 4.1 Formulas for the maximum deflection of beams in accordance with different boundary layer conditions.

4.3 STRENGTH DESIGN

The bending moment and shear force are borne predominantly by the flanges and the webs, respectively, in the early design. The formula used for isotropic materials may be used to approximate the bending stress:

$$\sigma = \frac{M_c}{I} \tag{4.2}$$

where M denotes the bending moment, I denotes the moment of inertia, and C (from the neutral axis to the top or bottom of the section) denotes the section's height or width. Because the tensile and compressive strengths of the laminate differ, it is necessary to verify both positive and negative values of stress for composite beams (rather than just the biggest absolute value).

The needed section modulus is derived by defining the section modulus and substituting the stress in Eq. (4.2) with the factored strength value F_{xt} or F_{xc} (tensile or compressive). The required modulus is:

$$Z = \frac{\alpha M}{\phi F_x}, \quad Z = \frac{I}{C}. \tag{4.3}$$

where α is the load factor. The condition of stress in the flanges for the majority of commonly used cross-sections (I-beams, box-beams, etc.) is primarily axial stress x. The strength F_x is therefore either the laminate's compressive (F_{xc}) or tensile (F_{xt}) strength, depending on the sign of the stress. From carpet plots, one may infer a laminate's strength. On the other hand, the axial stress changes sign across the thickness of the laminate, and the flexural strength F_{bx} of the laminate should be employed instead for rectangular solid beams with the laminae perpendicular to the load direction.

The shear stress in the web is determined using Jourawski's formula [1] for isotropic materials:

$$T = \frac{QV}{It} \tag{4.4}$$

where V is the shear force determined by the shear diagram, I is the beam's moment of inertia, t is the web's thickness, and Q is the initial moment of area. The neutral axis experiences the most shear stress since Q is greatest there. The laminate in the webs, which has a laminate shear strength F_{xy} specified in the carpet plot, must resist the shear force. Consequently, the design must meet:

$$\phi F_{xy} > \frac{\alpha QV}{It} \tag{4.5}$$

4.4 BUCKLING DESIGN

Compressive stresses are introduced into some areas of the cross-section of a thin-walled beam when it is bent, often at one of the flanges. That area of the cross-section might buckle because of compressive stresses.

The laminate's material characteristics, geometry, and the way distinct cross-sectional parts are supported all affect how resistant the laminate is to buckle. The buckling load per unit width is given as:

$$N_x^{CR} = \frac{2\pi^2}{b^2}\left(\sqrt{D_{11}D_{22}} + D_{12} + 2D_{66}\right) \tag{4.6}$$

The compressive load (per unit width of flange) may be calculated by adding forces and moments under the assumption that the two flanges of the box beam are entirely carrying the bending moment.

$$N_x = \frac{M}{bh} \tag{4.7}$$

where b is the width of both flanges and h is the height of the beam. The calculated compressive load width shouldn't be more than the limit load specified by Eq. (4.6). Because it assumes a simple support at the connections of the flange to the webs, Eq. (4.6) offers a conservative estimate of the buckling load of the flange. The flanges and webs are more or less tightly joined in actual sections. As a result, the rotation at the compression flange's borders is slightly limited. The margins of the flange are nearly clamped in the extreme case of extremely thick webs. The buckling load in this instance may be forecast using:

$$N_x^{CR} = \frac{\pi^2}{b^2}\left(4.6\sqrt{D_{11}D_{22}} + 2.67D_{12} + 5.33D_{66}\right) \tag{4.8}$$

The actual buckling load of the flanges lies between the values predicted by Eq. (4.6), because the compression flange is neither simply supported nor clamped (Eq. 4.8). Each flange of an I-beam may be depicted as two plates that are simply supported by the web, leaving the other edge free. After that, each half flange's buckling load may be anticipated using:

$$N_x^{CR} = \frac{12D_{66}}{b^2} + \frac{\pi^2 D_{11}}{a^2} \tag{4.9}$$

where a is the unsupported length of the flange or the area where the compressive load N_x acts and b is half of the flange width. The web's rotational constraints mean that the actual flange buckling stress is higher than that which was anticipated by Eq. (4.9). By fastening the flange to a panel, buckling of the flange may be avoided. In civil engineering construction, this is often done when the beam supports a floor system or a bridge deck. The true buckling strength of a thin-walled beam may only be estimated using the buckling formulae described in this section. Additional forms of failure that need to be considered during design include lateral-torsional and lateral-distortional buckling of beams [2].

4.5 BEHAVIOR OF A COLUMN

One of the several types of failure for a thin-walled column loaded axially under compression is discussed below.

4.5.1 EULER BUCKLING

A thin column that is axially compressed buckles with a lateral deflection that resembles a beam bending. It's known as Euler buckling. For a long, thin column, buckling happens at axial stress levels far lower than the material's compressive strength. The cross-sectional shape does not change when the column sags with lateral displacement. The critical load (or buckling load) of a long column depends on its length, bending stiffness (EI), and the type of support used at its ends:

$$P_{CR} = \frac{EI \ (\pi^2)}{L_e^2} \tag{4.10}$$

L_e is the length that can be achieved by using the actual length L and the coefficient of constraint, and it is defined as:

$$L_e = KL \tag{4.11}$$

It is possible to theoretically calculate the coefficient K or to modify it to match experimental results [3,4] (Table 4.1).

The same bending stiffness (EI) is used to forecast beam deflections. Using carpet plots, choosing a value for E_x, and creating the shape to provide the necessary moment of inertia "I," it is possible to calculate (EI) during early design. Bracing is typically used in practice to avoid Euler buckling. To increase the buckling load until the mode of failure switches to local buckling or material failure, the column is braced to shorten its effective length.

4.5.2 LOCAL BUCKLING

Thin-walled columns frequently exhibit buckling of the walls without overall Euler buckling deflection. In this instance, the cross-sectional form changes but the column's axis stays straight. Using the calculations in the preceding sections, it is possible to analyze each wall separately for buckling; however, the loads

TABLE 4.1
Values of K for Different End Restraint

End-restraint	K_{Theory}	K_{Steel}	K_{Wood}
Pinned-pinned	1.00	1.00	1.00
Clamped-clamped	0.50	0.65	0.65
Pinned-clamped	0.70	0.80	0.80
Clamped-free	2.00	2.10	2.40

projected in this method may contain large inaccuracies. To get correct findings, the analysis should consider the entire cross-section, and the finite element method is frequently utilized for this.

4.5.3 MATERIAL ERROR

Only when the cross-sectional walls of the material are thick, and narrow is the material's compressive strength obtained. In this situation, the local buckling stress may be greater than the material's compressive strength. A laminate stress analysis and failure evaluation may be used to anticipate material failure by simply comparing the applied compressive stress to the laminate F_{xc}'s compressive strength.

4.5.4 MODE INTERACTION

Any of the aforementioned nodes in combination can cause the column to fail. For instance, for pultruded wide flange I-beams of intermediate length, the local and Euler nodes interact. The critical load in this situation can be calculated using an empirical interaction factor [5]:

$$\frac{P_{CR}}{P_L} = \frac{1+1/\lambda^2}{2c} - \sqrt{\left(\frac{1+\frac{1}{\lambda^2}}{2c}\right)^2 - \frac{1}{c\lambda^2}} \qquad (4.12)$$

where λ is:

$$\lambda \approx KL\sqrt{\frac{P_L}{\pi^2 EI}} \qquad (4.13)$$

where P_L is the local buckling load, c is the interaction coefficient, K is the restraint coefficient, and EI is the bending stiffness.

REFERENCES

[1] J. M. Gere, and W. C. Young, *Advanced Mechanics of Materials*, Macmillan, New York, 1985.

[2] E. J. Barbero, and I. G. Raftoyiannis, "Lateral and distortional buckling of pultruded I-beams," *Compos. Struct.*, 27, 261–268 (1994).

[3] J. E. Shigley, and C. R. Mischke, *Mechanical Engineering Design*, Fifth Edition, McGraw Hill, New York, 1989.

[4] J. C. Smith, *Structural Analysis*, Harper and Row, New York, 1988.

[5] E. J. Barbero, and J. Tomblin, "A phenomenological design equation for FRP columns with interaction between local and global buckling," *Thin Walled Struct.*, 18, 117–131 (1994).

5 Stiffened Panels and Plates

5.1 BENDING OF A PLATE

On carpet plots of laminates' flexural strength, the preliminary design of laminates for bending is based. For isotropic materials, the following formulae can be used to calculate the value of stress at failure on the plate surface:

$$F_x^b = \frac{6M_x}{t^2} \tag{5.1}$$

$$F_y^b = \frac{6M_y}{t^2} \tag{5.2}$$

$$F_{xy}^b = \frac{6M_{xy}}{t^2} \tag{5.3}$$

where M_x, M_y, and M_{xy} are the bending moment values that lead to the laminate's actual failure. Suppliers of materials occasionally report flexural strength numbers based on experimental data for a select group of laminate configurations, such as unidirectional and quasi-isotropic. Flexural strength experimental data may be calculated using ASTM D 790. Depending on the design objectives, the values may be based on either first-ply failure (FPF) or last-ply failure (LPF). When just one bending moment is applied at a time, failure of the real laminate is expected. The strength ratio for last-ply failure, for instance, is the value of M_x that ultimately results in the failure of the laminate when applying a unit moment $M_{x=1}$. The flexural strength F_{bx} is then calculated using Eq. (5.1).

5.2 BUCKLING OF A PLATE

When a structure or a portion of a structure is subjected to compressive loads, buckling occurs. The structural collapse that comes from buckling is caused by substantial deflections and a drop in the structure's stiffness compared to the unbuckled condition. The intricate subject of composite material buckles is outside the purview of this work. However, a few specific circumstances can be precisely resolved, or straightforward approximation formulae might be suggested.

 DOI: 10.1201/9781003429197-5

The solutions in this section are for orthotropic laminates that are symmetric, with $[B]=[0]$ and $D_{16}=D_{26}=0$. The equations may predict incorrect outcomes for balanced symmetric laminates with nonzero values for D_{16} and D_{26}. The size of D_{16} and D_{26} have an impact on the error's magnitude. The off-axis laminae should be positioned in pairs ($\pm\theta$) and as near to the center surface as feasible to reduce these coupling coefficients.

To lessen the coupling effects, utilize balanced, biaxial woven or stitched carpets. The effects of transverse shear deformation, which are crucial for buckling, weren't considered when the equations in this section were derived. Therefore, unless the plate's dimensions (a and b) are significantly bigger than the thickness (a, $b>20t$), these equations are not conservative.

5.2.1 SIMPLY SUPPORTED ON ALL THE EDGES

The buckling load per unit length is in the case of a rectangular plate that is simply supported along the boundary and is subject to one edge load:

$$N_x^{CR} = \frac{\pi^2 D_{22}}{b^2}\left[m^2 \frac{D_{11}}{D_{22}}\left(\frac{b}{a}\right)^2 + 2\frac{D_{12}+2D_{66}}{D_{22}} + \frac{1}{m^2}\left(\frac{a}{b}\right)^2 \right] \quad (5.4)$$

where m is the number of waves in the loading direction that make up the buckled form (for instance, $m=2$). For this kind of loading and boundary conditions, there are $n=1$ waves. The closest integer to the real number R_m can be used to calculate the number m.

$$R_m = \left(\frac{a}{b}\right)\left(\frac{D_{22}}{D_{11}}\right)^{0.25} \quad (5.5)$$

If ($a>b$), the plate is long:

$$N_x^{CR} = \frac{2\pi^2}{b^2}\left(\sqrt{D_{11}D_{22}} + D_{12} + 2D_{66} \right) \quad (5.6)$$

Equation (5.4) also provides a good approximation when the loaded edges are clamped, where clamped means that the rotation and transverse deflection are restricted but the in-plane displacements are free. In the preliminary design, the coefficients of the bending stiffness matrix [D] can be computed using laminate bending moduli given in carpet plots:

$$\sigma_x^{CR} = \frac{\pi^2 E_y^b}{12\Delta}\left(\frac{t}{b}\right)^2\left[m^2 \frac{E_x^b}{E_y^b}\left(\frac{b}{a}\right)^2 + 2\frac{E_y^b v_{xy}^b + 2G_{xy}^b\Delta}{E_y^b} + \frac{1}{m^2}\left(\frac{a}{b}\right)^2 \right] \quad (5.7)$$

$$R_m = \left(\frac{a}{b}\right)\left(\frac{E_y^b}{E_x^b}\right)^{0.25} \quad (5.8)$$

$$\Delta = 1 - v_{xy}^b v_{yx}^b; \quad v_{yx}^b = \frac{E_y^b}{E_x^b} v_{xy}^b \tag{5.9}$$

$$\sigma_x^{CR} = \frac{N_x^{CR}}{t} = \left(\frac{t}{b}\right)^2 \frac{\pi^2}{6\Delta} \left(\sqrt{E_x^b E_y^b} + E_y^b v_{xy}^b + 2G_{xy}^b \Delta \right) \tag{5.10}$$

Take note of how the number of modes, m, grows as the plate lengthens. Given by Eq. (5.10), the loci of minimum values are a common estimate for design that is cautious. The usage of sandwich plates is encouraged by the fact that the buckling strength in Eq. (5.10) increases as a square of thickness. It is also clear that the ratio (t/b) is a key factor in the plate's buckling strength. Therefore, reducing the effective width of the plate by intermediate supports is the easiest approach to increase the buckling load.

To do this, longitudinal stringers are added to support the plate at various spots throughout its breadth. The geometric characteristics, such as (t/b) 2 in Eq. (5.7), often have a greater impact than the material parameters (α, β, γ) and the order of the laminate stacking.

As a result, the geometry dominates the buckling response. For a given material system, empirical equations based on experimental data or parametric studies have been proposed for the preliminary design of compression members based on material stiffness (carbon vs. E-glass). Experiments confirm that geometric parameters dominate buckling behavior for a given material system.

5.2.2 CLAMPED ON ALL SIDES

The buckling load per unit width for a long plate ($a/b > 4$) loaded exclusively in the x-direction may be calculated by:

$$N_x^{CR} = \frac{\pi^2}{b^2}\left(4.6\sqrt{D_{11}D_{22}} + 2.67D_{12} + 5.33D_{66}\right) \tag{5.11}$$

The buckling stress is:

$$\sigma_x^{CR} = \frac{N_x^{CR}}{t} = \left(\frac{t}{b}\right)^2 \frac{\pi^2}{12\Delta}\left(4.6\sqrt{E_x^b E_y^b} + 2.67E_y^b v_{xy}^b + 5.33G_{xy}^b \Delta\right) \tag{5.12}$$

5.2.3 SINGLE FREE EDGE

A long plate with length a and width b ($a/b > 4$), three readily supported edges, one unloaded edge free, and loaded along length, has the buckling load per unit width given by:

$$N_x^{CR} = \frac{12D_{66}}{b^2} + \frac{\pi^2 D_{11}}{a^2} \tag{5.13}$$

The buckling stress is:

$$\sigma_x^{CR} = \frac{N_x^{CR}}{t} = \left(\frac{t}{b}\right)^2 \left(G_{xy}^b + \pi^2 \left(\frac{b}{a}\right)^2 \frac{E_x^b}{12\Delta}\right) \tag{5.14}$$

5.2.4 BI-AXIAL LOADING

The buckling load of a rectangular plate ($1 \leq a/b < \infty$) with all edges simply supported and subject to loads N_x and $N_x = kN_x$, with k constant, is determined by the minimum with respect to the two loads m and n:

$$N_x^{CR} = \frac{\pi^2}{b^2} \frac{D_{11}m^4c^4 + 2Hm^2n^2c^2 + D_{22}n^4}{m^2c^2 + kn^2}$$

$$C = b / a \tag{5.15}$$

$$H = D_{12} + 2D_{66}$$

$$k = \frac{N_y}{N_x} = \frac{N_y^{CR}}{N_x^{CR}} \tag{5.16}$$

By attempting all possible combinations of the integers m and n, which, respectively, denote the number of half-waves in the x and y directions, the minimal value of (5.16) is discovered.

5.2.5 FIXED UN-LOADED EDGES

A rectangular plate ($1 \leq a/b < \infty$) loaded exclusively in the x-direction is an example. The buckling load is determined by minimizing the following equation with respect to m if the loaded corners are simply supported for the unloaded sides they are clamped:

$$N_x^{CR} = \frac{\pi^2}{b^2} \left[D_{11}m^2c^2 + 2.67D_{12} + 5.33\frac{D_{22}}{c^2} + D_{66} + \frac{1}{m^2} \right] \tag{5.17}$$

where m is the number of half-waves and $c = b/a$.

5.3 STIFFENED PANELS

A flat or curved laminate is used to make stiffened panels, which are then strengthened by a grid of longitudinal and transverse stiffeners. Greater buckling resistance under in-plane stresses and increased bending stiffness under transverse and bending loads are both provided by the stiffeners.

Depending on the use, the primary constituents of stiffened panels go by different names. The skin is reinforced and held up by the ribs and stringers.

Although the transverse reinforcements are known as frames, the longitudinal reinforcements along the fuselage are also known as stringers. In the construction of ships, longitudinal reinforcements along the hull and bulkheads in the transverse direction are used. A deck is the term used to describe the outer layer of a bridge that is supported by cross-beams and stringers. In floor systems, transverse beams and longitudinal joists support the floor deck. Purlings and rafters are used in roof building to support the roof panels. Function, rigidity, etc. are factors that affect how ribs, frames, bulkheads, etc. differ. It is necessary to understand that each component's preliminary design can be completed individually for the purposes of this section. Whether or not the main load is bending or moving in-plane determines the design strategy. The fundamental issue with in-plane load is buckling. Deflections and strength are the primary factors to take into account while bending; however, lateral-torsional buckling of the stringers may also be a problem. It is outside the scope of this book to address stiffened panel design in detail. As a result, just the fundamental elements of the design will be discussed, with an emphasis on the specifics of the usage of composite materials.

5.3.1 STIFFENED PANELS UNDER LOADS WITH BENDING

As was previously mentioned, a thorough investigation of the bending of plates, whether stiffened or not, is challenging. By keeping in mind that the spacing between stringers and ribs (such as those on frames and bulkheads) is typically considerably narrower, the design may be made simpler.

5.3.1.1 Design of the Skin

The skin's design is predicated on the idea that the stringer spacing—typically 6/20 for an aircraft fuselage—is significantly less than the rib or frame spacing. Under these circumstances, the transverse (across the stringers) bending moment of the skin is significantly greater than the longitudinal (along the stringers) bending moment. As a result, the skin may be created by separating a thin transverse strip and considering it as a beam with dimensions b, t, and l that are equal to the spacing between the stringers l_s. The torsional stiffness of the stringers affects the support circumstances at the end of the beam. The skin can be thought of as a clamped beam if the stringers are exceedingly strong and do not twist.

The analysis of the skin as a continuous beam is possible if the stringers are not stiff in torsion. The load on the beam is $q_0 = p_0 b$, assuming a uniformly distributed pressure of p_0 on the skin. For the clamped situation, the maximum moment and deflection of the skin are:

$$M = -\frac{q_0 l_s^2}{12} \left(\text{at supports} \right) \tag{5.18}$$

$$M = \frac{q_0 l_s^2}{24} \left(\text{at mid-span} \right) \tag{5.19}$$

$$\delta = -\frac{q_0 l_s^2}{384EI} \tag{5.20}$$

$$M \cong \pm\frac{q_0 l_s^2}{10} \tag{5.21}$$

$$I = \frac{bt^3}{12} \tag{5.22}$$

After choosing a laminate and an E_x value, Eqs. (5.20) and (5.22) provide thickness sizing if the design is deflection controlled. After choosing a laminate and a flexural strength F_{bx} value, Eqs. (5.18) and (5.21) allow for the size of the thickness of the skin if the design is strength regulated. The design should take appropriate load and resistance considerations into account. The design procedure results in the selection of laminates reinforced largely in the x-direction (1) because the bending moment M_y in the skin has been disregarded. Such a decision would be wrong since it is likely that substantial values of the bending moment M_y will be found close to the ribs. Additionally, the intersection of the rib and stringer is likely to include high values of the twisting moment M_{xy} and the shear force N_{xy}. The designer is thus recommended to stay away from laminates that are mostly unidirectional. It should be emphasized that this streamlined design approach has mostly been used with quasi-isotropic laminates, laminates reinforced with woven fabrics, or reinforcements made of continuous or chopped strand mats.

5.3.1.2 Design of the Stringer

The stringer design may be carried out utilizing the tributary area approach, in line with the skin design. The load delivered to the skin midway between the two neighboring stiffeners, known as the tributary region, is what is taken by the stringer. The stringer is then created as a beam with a load that is equally distributed across it, $q_0 = p_0 l_t$. The tributary width is equal to the stringer spacing if the stringer spacing is consistent (or pitch l_s). The cross-members' rigidity affects the support conditions. The stringer is modeled as a clamped beam with the maximum moment and deflection determined by Eq. (5.21) since frames and bulkheads are often extremely stiff Eq. (5.22).

A balance between fabrication, cost, torsional and bending stiffness, and other factors unique to the application at hand led to the selection of the stringer design. Several stiffener geometries are shown. The stiffener is constructed as a beam once the shape has been decided upon in order to offer sufficient strength and adhere to any deflection constraints. The principal pressures are oriented along the stiffener's length, making it function as a beam. Therefore, the in-plane characteristics E_x and F_x for bending and G_{xy} and F_{xy} for shear are employed as the apparent properties. Typically, a piece of the skin is added to the stringer to account for the skin's contribution to the stiffness and strength in the direction of the stringer.

However, due to uneven shear distribution (shear lag), a poor skin-stringer connection (loss of composite action), or skin buckling, not all of the tributary width fully contributes to the stringer. As a result, the stringer merely receives the skin's effective width.

5.3.1.3 Under In-Plane Loads, a Stiffened Panel

When a stiffened panel is subjected to intense edge loads (N_x, N_y) or shear loads (N_{xy}), a variety of failure modes may manifest, although buckling is often what governs the design. Numerous buckling modes are possible. Overall or global buckling happens when the entire panel collapses in a way that deforms it as a whole. Skin buckling happens when the stiffeners maintain their original shape while the skin only buckles. The panel could still be able to support the imposed loads in this situation. Open-section stiffeners are more susceptible to torsional buckling of the stiffeners (blade, J, I, etc.). This type of buckling severely weakens the panel and has the potential to cause widespread buckling and collapse.

Close-section stiffeners, like the hat stiffener, provide strong torsional stiffness that both prevents the stiffeners from bowing in one direction and adds to the panel's overall stability [3].

Last but not least, local buckling of the stiffener happens when the stiffener's walls (flange or web) buckle locally. Localized stiffener buckling weakens the panel and can cause structural collapse. Close-section stiffeners can also buckle locally, although open-section stiffeners are more prone to it, especially on walls with free edges. Closed-section stiffeners may utilize a filler (foam or honeycomb) to minimize local buckling, but this must take into account the filler's potential for water retention. The design of a stiffened panel is influenced by a variety of practical factors, such as manufacturing costs and splicing simplicity. As a result, there is no attempt in this section to give a thorough analysis of panel design. Instead, a succinct explanation of the most basic design approach is provided to show how composite materials fit into the traditional panel design methodologies. Using design-optimization systems based on semi-analytical techniques, better forecasts of buckling loads and post-buckling behavior may be made. These programs stand for a compromise between the drawbacks of the here described closed-form solutions and the complexity and cost of computing involved in nonlinear finite element analysis. These software programs can be used to directly construct a panel or to evaluate a specific panel in great detail. When a novel circumstance develops, a preliminary stiffened panel design may be carried out utilizing straightforward mathematics to disclose the key characteristics of several design choices, including stiffener spacing. The design can then be revised after a thorough examination with a software program reveals a more accurate estimate of the panel's genuine load-bearing capacity. In the end, prototype testing will probably be utilized to confirm the panel's performance.

5.3.1.4 Stiffener and Skin Strength

The panel's whole cross-section is subjected to the compressive edge load. According to the indifferent axial stiffness, the load is distributed proportionately to the skin and longitudinal stiffeners (EA). Given that the skin and stiffener both experience the same in-plane strain, the load borne by each component may be calculated. In order to estimate the in-plane strain, edge load, and applied load at the neutral surface of the panel, Poisson effects are ignored as:

$$\varepsilon_x = \frac{N_x l_s}{EA} \quad (5.23)$$

where (EA) is the axial stiffness of a piece of the panel including a stiffener and a portion of the skin of width l_s, and l_s is the stiffener pitch. You may calculate the value of (EA) by modeling the panel's section as a thin-walled beam. The axial stiffness is determined by the laminate moduli E_s^{sk} of the skin and E_x^s of the stiffener:

$$EA = E_s^{sk} A_{sk} + E_x^s A_s \quad (5.24)$$

$A_{sk} = t_{sk} l_s$ and A_s are the areas of the skin and the stiffener.

The region is made to support the load utilizing the laminate's in-plane strength F_{xc}, so long as there is no buckling. If the projected strain ε_x is less than the compressive strain at which the laminate will fail $(\varepsilon_{xc} = F_{xc}/E_x)$, the material won't fail under compression. Since the strain in the skin and the stiffener is the same in terms of stress, the load on each component equals:

$$P_s = E_x^s A_s \varepsilon_x$$
$$N_x^{sk} = E_x^{sk} t_{sk} \varepsilon_x \quad (5.25)$$

where A_s is the area of the stiffener and t_{sk} is the thickness of the skin, and P_s is the load in the stiffener, N_x^{sk} is the load per unit width (N/m) of the skin, E_x^{sk} and E_x^s are the laminate moduli of skin and stiffener, respectively. With this knowledge, it is simple to assess the skin's and stiffener's material compressive strength. In a laminate analysis software, N_x^{sk} may be used to analyze the laminate for the skin using a failure criterion. Otherwise, if the material's compressive strength is not attained as long as $N_x^{sk}/t_{sk} < F_{xc}$, the laminate's in-plane strength F_{xc} can be obtained from a carpet plot. Similar analysis may be performed on the stiffener. However, the majority of useful designs will buckle rather than fail due to material strength.

5.3.1.5 Buckling of a Stiffened Panel

The critical buckling stress grows with the square of the ratio t/b for a singly supported plate of width b, as illustrated in Eq. (5.9). Therefore, cutting the plate into smaller segments can lower the width b, increasing the laminate's buckling strength. The smaller plates can be supported adequately by the use of longitudinal stiffeners, also known as stringers. The longitudinal stiffeners can buckle as columns and bear some of the axial strain. By including transverse stiffeners (such as ribs, frames, or bulkheads), their effective length can be decreased while the column buckling stress is raised. If the panel's sides are not supported, the entire panel might collapse as a column under an in-plane load of N_x. The panel's buckling load provided by Eq. (5.9) is:

$$N_x^{CR} = \frac{1}{l_s} \frac{\pi^2 (EI)}{L_e^2} \tag{5.26}$$

where L_e is the effective length, described in Eq. (5.10), and E is the bending stiffness (relative to the neutral surface of bending) of a representative area of the panel made up of one stiffener and a section of skin with a width equal to the stiffener spacing l_s. The panel has a unique value of laminate in-plane moduli if the skin and stiffeners are built from the same material configuration (identical values of α, β, γ) E_x. The buckling load per unit width is as a result of:

$$N_x^{CR} = \frac{1}{l_s} \frac{\pi^2 E_x I}{L_e^2} \tag{5.27}$$

where I is the moment of inertia of the panel's width l_s part with respect to the neutral surface of bending, and L_e is the column's effective length, which varies on the actual length and the boundary conditions Eq. (5.10). The method for thin-walled beams may be used to compute the bending stiffness (EI) for a stiffened panel with any material configuration [2]. The critical strain at which Euler buckling occurs is then easily calculated using the formula:

$$\varepsilon_x^{CR} = \frac{N_x^{CR}}{EA} \tag{5.28}$$

where (EA) represents the panel's examined section's axial stiffness (5.25). The panel functions as a plate if it is supported on both sides. The equations in Section 5.2 may be used to calculate the panel's total buckling load as long as the bending stiffness coefficients are changed to equivalent values for the panel. By averaging the contribution of the stiffeners to the plate, these comparable plate stiffnesses are derived. Individual plate components in the skin or the stiffeners' walls may buckle as plates in addition to overall buckling (column or plate buckling) (long plates). The equations from Section 5.2 can be used to calculate the buckling loads of these plate components. According to the edge supports, it is possible to evaluate the skin portions that are constrained by longitudinal and transverse stiffeners as simply supported plates and compute the local buckling load of the stiffener walls using Eqs. (5.6) and (5.13). The approach entails breaking apart the panel and stiffeners into a number of small flat plates and examining each plate separately. The cross-section, for instance, has been separated into sections that represent flat plates to be examined independently. The estimated buckling load N_x^{CR} of plates supported at both sides (such as segments 2, 3, and 4) is given by Eq. (5.6). For segments having a free edge, such segments 1 and 5, Eq. (5.13) is utilized. Last but not least, the section of skin supported on both borders between two stiffeners is examined (5.6). The skin's bonded section to the stiffener is considered to be a part of the stiffener, the buckling strain is,

$$\varepsilon_x^{CR} = \frac{N_x^{CR}}{tE_x} \tag{5.29}$$

where t and E_x are the element's thickness and laminate modulus under consideration. Incipient buckling is indicated by the minimal value of ε_x^{CR} obtained using Eqs. (5.28) and (5.29). Simply put, the equivalent load per unit width of the panel:

$$N_x^{CR} = \frac{1}{l_s}(EA)\varepsilon_x^{CR} \tag{5.30}$$

where (EA) is the panel's axial stiffness along the l_s width region. After buckling, both the stiffener and the skin may support further loads; this process is known as post-buckling.

5.3.1.6 Crippling and Post-buckling

Many times laminated composite panels can withstand loads greater than the buckling loads. The post-buckling range stretches from the initial buckling load up to the load that results in material failure or structural collapse. The deformations (mostly the transverse deflections) increase dramatically during post-buckling, as does the buckling shape or mode. Large stresses are ultimately responsible for material failure, which results in collapse. Crippling strength is the name for the collapse load. It needs both a solid model of material deterioration and numerical methods that can follow the post-buckling route to provide accurate predictions of crippling strength values.

The two unloaded edges were merely supported for the no-edge-free situation. In the situation of one unloaded edge being free, the other edge was just supported. The loaded edges were repaired in both situations. To create crippling curves that are applicable for various laminates, the data are normalized by the laminate's compressive strength (F_{xc}) [1]. This normalization method is effective for laminates that have more than 25% of 45° and more than 25% of 0° laminae, but it is ineffective for laminates that contain 100% of ±45°, 100% of 0°, or 100% of 0°/90° laminae. By drawing a line parallel to the data's least square fit and ensuring that all data points lie above this line, it is possible to derive a lower limit for the data, as illustrated in the figures. Based on the graph, the width/thickness ratio (b/t) of the wall, its boundary conditions (one edge-free or no-edge-free), and the compressive strength of the laminate, the crippling strength F_{cc} of each wall (or segment) of the stiffener is determined (F_{xc}). The geometric parameter b/t dominates the crippling behavior relative to the laminate structure for the class of laminates under consideration.

This does not suggest that laminate construction has no influence on the crippling strength F_{cc}, but rather that the laminate compressive strength F_{xc} may account for the effect of laminate construction.

$$\varepsilon_{cc} = \alpha \left(\frac{\varepsilon_{xc}}{\varepsilon_x^{CR}} \right)^\beta \varepsilon_x^{CR} \tag{5.31}$$

where ε_{xc} is the plate element's strain to failure in compression ($\varepsilon_{xc} = F_{xc}/E_x$), α and β are empirical constants stated in study [1], and ε_x^{CR} is the buckling strain.

It is still necessary to compute the crippling load N_x^{CC} of the panel once the crippling strain cc or crippling strength F_{cc} of each wall has been determined. Two cases must be taken into account. The panel collapses if the wall with the lowest crippling strain is a typical wall to the panel, such as a blade stiffener. The stiffener's load in this instance is:

$$P_s = E_x^s A_s \varepsilon_{cc} \tag{5.32}$$

The crippled plate element will continue to support weight until a second wall cripples with strain ε_{cc2}, which can happen if the wall with the lowest crippling strain, ε_{cc1}, is parallel to the panel, like the top of a hat stiffener. Consequently, the stiffener's load is:

$$P_s = E_x^s b_1 t_1 \left(\varepsilon_{cc1} - \varepsilon_{cc2} \right) + \varepsilon_{cc2} E_x^s A_s \tag{5.33}$$

where E_x^s and A_s are the stiffener's modulus and area, respectively, and b_1, t_1, and ε_{cc1} are the first wall's width, thickness, and crippling strain; ε_{cc2} is the second wall's crippling strain. When the skin is extremely thin, it may give way before the stiffeners do. In this situation, the skin may be completely disregarded or a smaller "effective width" may be employed to symbolize the skin's diminished contribution. Using a well-known metal design formula, the effective width of skin was calculated as

$$w_e = 1.9 t_{sk} \left(\varepsilon_x \right)^{-0.5} \tag{5.34}$$

$$P_{sk} = 1.9 t_{sk}^2 E_x^{sk} \left(\varepsilon_{cc} \right)^{0.5} \tag{5.35}$$

where, depending on the circumstance, ε_{cc} is either equal to ε_{cc2} of (5.33) or the crippling strain of the stiffener utilized in Eq. (5.32). Unless the stiffeners are sewn, riveted, or otherwise securely fastened to the skin, most composite stiffened panels fail when the skin separates from the stiffeners. The calculation is based on the presumption that when the panel strain approaches the crippling strain of the skin, the skin will separate from the stiffener. The skin-damaging strain was calculated using Eq. (5.32) in the case of a single edge. The failure load resulting from skin detachment is thus:

$$N_x^{ss} = \varepsilon_{ss}\left(\frac{E_x^s A_s}{l_s} + E_x^{sk} t_{sk}\right) \qquad (5.36)$$

when experimental failure loads and failure strains (ε_F) are provided for a variety of stiffeners and stiffened panels, the approach is used to forecast the failure load. ε_F is the experimental strain at failure, and ε_{xc} is the laminate's inherent compressive strain at failure in both pictures (without buckling effects).

REFERENCES

[1] MIL-HDBK-17-3 D, *Military Handbook, Polymer Matrix Composites, Vol. 3*, Utilization of Data, Department of Defense, Washington, DC (1994).
[2] M. S. Troitsky, Stiffened Plates-Bending, *Stability and Vibrations*, Elseiver, New York (1976).
[3] Gibbs and Cox, Inc., *Marine Design Manual for Fiber-Glass Reinforced Plastic*, McGraw Hill, New York (1960).

6 Effects on the Environment, Fatigue, and Performance of Fiber Composites

6.1 FATIGUE

Although the maximum stress never surpasses the material's ultimate static strength, it is well known that materials can fail when subjected to repeated fluctuating or alternating pressures. A material's strength is reduced by load cycling, or its fatigue strength is lower than its static strength, to put it another way. The vast majority of materials used today, such as metals, polymers, and composites, come within this category. In the field, fatigue loads are practically never averted. Because of this, static strength and fatigue analysis are now both essential design criteria for modern designs. Fatigue analysis is a significant concern in the transportation industry, notably in aircraft, because there is a requirement for increased structural material performance.

According to this perspective, several researchers have looked into the fatigue of composite materials. Although clear design requirements, similar to those for metal fatigue, have not yet been established, many significant elements of composite fatigue are now well recognized.

It is well known that unidirectional continuous-fiber-reinforced composites have exceptional fatigue resistance in the direction of the fiber. This is because the load in a unidirectional composite is predominantly borne by the fibers, which typically display superior fatigue resistance. However, composites are primarily utilized as laminates in real buildings.

Due to variances in ply orientation, certain plies are physically damaged well before they fracture, making them weaker than others in the direction of loading. There are many ways that damage might appear, including as failure of the fiber-matrix interface, matrix cracking or crazing, fiber breakage, and void growth (i.e., separation of plies or delamination). Since it soon leads to final fracture, the formation of visible damage (such as a crack) in metals is frequently viewed as harmful. However, this is not necessarily the case with composite materials since, although the initial damage may appear early in the fatigue life, it may be halted by the internal structure of the composite.

The fact that design loads in crucial applications should be lower than what is necessary to cause any damage to the composite should be emphasized.

DOI: 10.1201/9781003429197-6

Typically, damage to individual plies diminishes the laminate's elastic properties, which could eventually result in structural failure (e.g., excessive deformation). However, it's possible for this to happen before the laminate is in risk of rupturing. Because of this, failure in composite materials has different meanings depending on the application. Instead of requiring complete rupture, the failure criterion in an application where deformation or a change in stiffness must be limited may be a loss of stiffness equal to a predetermined percentage of the original stiffness.

The two requirements almost match up in this situation since metals don't alter significantly in stiffness unless significant breaking happens. These evident factors make it impossible to infer a good design method for composite materials used in fatigue applications from methods used with metals. A designer must exercise caution when using their own judgment when there are no established design methods in place. The design engineer, on the other hand, can unquestionably profit from a careful investigation of several aspects of composite fatigue behavior. This section's presentation reflects this point of view. We'll start by examining how fatigue damage develops and progresses, as well as how it impacts composite properties.

The effects of material properties including matrix material, ply orientation, fiber content, and fiber finish are then investigated, along with testing parameters like mean stress and frequency. The current trend in developing empirical relationships for evaluating fatigue damage and fatigue life is also briefly discussed. The remaining two parts cover fatigue of short-fiber composites as well as the behavior of high-modulus fiber-reinforced composites.

6.2 FATIGUE DAMAGE

The factors listed below result in fatigue damage. The failure modes under coupled corrosion, fatigue, or tensile loading are frequently localized, typically in terms of crack growth.

6.2.1 DAMAGE/CRACK INITIATION

On the damage start and propagation mechanisms in fatigue composite laminates, numerous studies have been done. It has been determined that the initial stage of the damage process is the separation of fibers from the matrix (known as debonding) in the fiber-rich portions of the plies where the fibers lay perpendicular or at a significant angle to the loading direction. High concentrations of stress and strain at the fiber matrix contact are what start these cracks. When a crack first starts, it frequently spreads across fibers, especially along the fiber matrix contact. A typical cross-ply crack is seen in study [1]. The crack extends the full width of the ply and is typically perpendicular to the direction of load.

If the applied stress exceeds the local ply strength during the first cycle of loading, cross-ply cracks may form. Depending on the laminate construction, this can happen at applied loads as low as 20% of the ultimate stress [2]. Cross-ply cracks occur more frequently as cycle count or stress increase. The cross-plies in study [1] show a variety of fracture types.

The initial damage manifests similarly in fiber composites with randomly oriented directions. A thin laminate made of chopped strand mat showed the first signs of degradation at around 30% of the expected ultimate tensile strength during a tensile test [3].

The initial damage in this instance was likewise caused by the strands that were parallel to the load line. Damage is not always associated with the ends of the strand and can start anywhere along its length. Strand debonding appears to be the damage's primary manifestation. Debonding fissures developing along fibers that are perpendicular to the path of tension, or at the largest angle to it, are hence the first sign of damage in a composite laminate.

6.2.2 Crack Arrest and Crack Branching

Cross-ply cracks spread across the entire width of the ply but are impossible to penetrate the next ply, particularly if the fibers of the neighboring ply are aligned in the direction of the stress. Cross-ply cracks thus stop at the junction of two plies. This element of fracture termination is clearly depicted in study [1]. On the other hand, the crack tip induces tension concentration beforehand. The interply contact provides the optimum location for a delamination crack due to the strong interlaminar strains that result. The delamination crack shown in the Figure 6.1 started close to the cross-ply crack's tip. There are more delamination fractures that manifest and spread as the number of cycles rises.

Delamination cracks appear concurrently with another type of deterioration. Additionally, the fibers in the longitudinal plies could start to break down and debond, which would cause cracks in the longitudinal plies. In the cross-section of the laminate in Figure 6.1, a longitudinal-ply crack is seen. Longitudinal-ply fractures do not have a certain path, as contrast to cross-ply cracks, which are often perpendicular to the load line.

6.2.3 Final Fracture

Later, when the composite has sufficiently deteriorated from longitudinal-ply cracks and delamination cracks, final breakage occurs. The longitudinal-ply cracks cause the longitudinal plies, which support the majority of the load, to weaken. Load distribution between plies is blocked by the existence of delamination cracks, which reduces the composite to a collection of isolated longitudinal plies working in tandem to maintain the imposed load. These longitudinal plies fail in succession when the weakest of them fails, which also affects the other longitudinal plies. The region close to the failure zone has obvious delamination, as seen in Figure 6.1. The delamination fractures, which cause the material's final fracture, are only clearly visible toward the end of the endurance test. For instance, Broutman and Sahu [2] were the first to make this observation after roughly 90% of the fatigue life, which was later confirmed by Dally and Agarwal [1].

6.2.4 INFLUENCE OF DAMAGE ON PROPERTIES

Internal cracking causes composite materials to become less rigid and robust. The development of cracks in a glass-epoxy cross-ply material has been associated by Broutman and Sahu [2] with changes in residual strength and modulus. Residual strength and stiffness decrease with increasing crack density. Alternately, two straight lines can be used to approximate the stress-strain curve of a virgin cross-ply material, giving the material two elastic moduli. Moduli can be classified as either primary or secondary. The material has a greater (primary) modulus since it is free of cracks before the test is initiated, and both the longitudinal and cross-plies equally contribute to the stiffness of the composite.

Cross-ply cracks develop as the load rises and the cross-plies' contribution to the composite stiffness declines, leading to a loss in modulus. In fatigue tests, longitudinal-ply cracks are the first to cause the modulus to drop, then cross-ply cracks, and finally delamination cracks. The material's stress-strain curve becomes linear as a result of the fatigue exposure, with a secondary modulus that is comparable to that of the virgin material. The modulus may fall below the secondary modulus of the material when longitudinal-ply cracks and delamination cracks appear as a result of fatigue loading. According to Dally and Agarwal [1], there is a quantifiable relationship between fracture density and modulus change for an E-glass-epoxy cross-ply laminate.

The average distance between two succeeding cracks in the cross-plies is known as the crack pitch. The material's static strength gradually declines as it undergoes an increasing number of cycles at a constant stress level. The first 25% of the fatigue life is when the majority of strength loss occurs, and after that point the rate of static strength loss reduces until the fatigue life is reached, and failure takes place. Again, the early strength loss is attributable to the failure of the cross-plies.

The loss of strength in the later stages of fatigue life is similarly gradual because longitudinal-ply cracks and delamination cracks form gradually. There is a sudden decrease of strength when the stronger plies fail in the last few fatigue cycles. Prior to this sudden loss of strength, individual plies start to deteriorate, but the reduction in overall strength is only slight. Curves like those in study [2] are typical of cross-plied materials. The fatigue of polyester resins reinforced with glass cloth was studied by Tanimoto and Amijima [4,5].

Their outcomes are strikingly similar to Broutman and Sahu's [2] as seen in study [2]. Additionally, they found that the residual interlaminar shear strength exhibits the same pattern as the residual tensile strength. In their investigation of the fatigue of glass-epoxy angle-ply laminates, Hahn and Kim [6] found that the material's secant modulus decreases with fatigue loading and that this decrease is associated with internal damage.

In addition to internal breaking damage, a rise in temperature causes a deterioration in the material's properties. Dally and Broutman [7] observed a significant temperature increase during the fatigue of a cross-ply material, particularly when the frequency is high. Glass-reinforced polypropylene underwent constant-deflection flexural testing, and Cessna et al. [8] utilized cycles to monitor load deterioration (proportional to modulus decay) (as shown in study [8]). It was also noted that viscoelastic energy dissipation, which is common in polymer-matrix composites, causes temperature to rise. The material becomes weaker and has a shorter fatigue life as a result of the temperature increase, which also signals growing fatigue damage. Cessna et al. [8] were able to extend by an order of magnitude both the cycles before the onset of stiffness change and the fracture life by cooling their specimens to maintain isothermal conditions.

6.3 EMPIRICAL RELATIONS FOR FATIGUE DAMAGED AND FATIGUE LIFE

A virtually infinite variety of laminates can be employed for structural applications. Once a laminate is chosen for use, experiments can be done to ascertain its fatigue characteristics. Of course, it would be impractical to approach the design issue from the opposite angle and evaluate every potential laminate in order to select the best one. However, it would be advantageous to create some straightforward equations for use in design analysis.

A framework to predict fatigue-induced strength loss was put forth by Broutman and Sahu [9]. According to their theory, loss of strength depends on the difference between static and fatigue strength as well as the ratio of applied-load cycles to expected fatigue life. This theory is useful in forecasting cumulative fatigue damage when the load cycle varies periodically, as it does in real constructions. However, as this hypothesis calls for the development of specific S-N curves for each material, it cannot be applied to design analysis. Mandell has connected the maximum stress-intensity factor and the rate of fracture propagation [10,11]. Additionally, no relationship between the fatigue strength and life has been established that might be applied in this situation for design purposes. The following relationship between fatigue strength and static strength of composite materials has been proposed by Hashin and Rotem [12]:

$$\sigma_f = \sigma_s f(R, N, n, \theta) \tag{6.1}$$

$f(R, N, n, \theta)$, where R stands for the stress ratio during fatigue cycling, N stands for the fatigue life, n stands for the frequency of load cycling, and θ stands for the fiber orientation for unidirectional composites, where σTr and σT are fatigue and static strengths, respectively. The function f, on the other hand, must be evaluated experimentally. As a result, its usefulness to design analysis is restricted. Equation (6.1) cannot be employed successfully until a simple way to evaluate this function can be found.

The S-N curves of composite materials have been reported to be frequently approximated by straight lines with the equation [13]:

$$\frac{\Delta S}{\sigma_u} = m \log N + b \tag{6.2}$$

The range of stress is ΔS, the ultimate tensile strength is denoted by σru, and the material constants are m and b. According to several studies [6,13], M and B could have values that are very near to 0.1 and 1.0, respectively. However, there isn't enough experimental evidence to suggest that using these values in design applications is a sure thing.

A power law is another useful connection for representing fatigue data:

$$N^k \Delta \epsilon = c \tag{6.3}$$

The strain range is $\Delta \epsilon$, and the material constants are k and c. This equation has been found to be highly useful for estimating metallic material fatigue life [14,15]. The value of c is connected to the ductility of the material, while k is known to fluctuate between 0.5 and 0.6 for most metals. This makes it possible to predict the fatigue behavior of metals based on their static properties. Even if the findings of Agarwal and Dally [13] and Hahn and Kim [6] agree with Eq. (6.3), the constants k and c for composite materials are not constants that apply to all materials. Furthermore, no connection has been shown between these constants and the static characteristics of materials. Eqs. (6.2) and (6.3) would become quite useful in design procedures if such a correlation were to be discovered in the future.

6.4 IMPACT

High-strain rate or impact loads are typical in many technological applications of composite materials. As a result, concerns about impact behavior or energy absorption as well as static strength considerations are used to determine a composite's suitability for such applications. Impact characteristics frequently deteriorate as a result of tensile quality improvement efforts. For instance, high-modulus fiber composites are more brittle than lower-modulus glass-fiber composites and put off less energy when they break.

Therefore, a full understanding of composite impact behavior is essential for designing safe and effective structures as well as for developing new composites with good impact and tensile properties. The discussion in this part is geared toward advancing that comprehension.

A material's impact behavior is greatly influenced by the fracture process caused by impact loads. Understanding the composites' fracture mechanism is therefore essential to comprehending how they behave under impact. The fracture process in composites is significantly more complicated than in homogeneous materials like metals or plastics because several microfailure events may take place during fracture propagation in a composite. It is believed that tiny, inherent faults in the material are what cause failure in fiber composites, just like failure in metals. Examples of flaws include brittle fibers, matrix faults, and deboned interfaces.

6.4.1 MECHANISMS FOR ABSORBING ENERGY AND FAILURE MODELS

A solid can absorb energy when it is loaded, whether it is static or dynamic. Two methods include material deformation and the production of new surfaces.

The first thing that happens is the material deforming. If there is enough energy in the loading, a fracture may start and spread, creating new surfaces. The second source of energy absorption is this one. Material deformation keeps going forward of the crack tip during fracture propagation. Only minimal deformation and minimal energy absorption occur in brittle materials like glass and ceramics. During fracture, more energy is absorbed as more surfaces are created. The quantity of energy needed to form a fracture surface is also little.

Brittle materials lack toughness or the ability to absorb large amounts of energy. Significant energy is absorbed during plastic deformations in ductile metals. This results in the high fracture toughness of ductile materials.

The discussion above makes it evident that a material's ability to absorb energy or toughness can be increased by enhancing either its plastic deformation capability or the creation of new surfaces after fracture. Metals frequently employ the first mechanism, and toughness can be increased via metallurgical procedures. By replacing a low-energy-absorbing component with a higher-energy-absorbing one, a composite's toughness can be increased. For instance, to create tougher hybrid composites, glass fibers are combined with carbon fiber composites.

In a composite with a fixed system of matrix material and reinforcing fibers, failure events occurring in the fracture process zone should be carefully analyzed and examined for changing material-impact behavior or toughness. The numerous failure events and related energies are discussed in the following paragraphs.

6.4.2 FIBER BREAKAGE

When a crack must spread in a direction normal to the fibers, fiber breaking will eventually occur, causing the laminate to separate completely. When a fiber's fracture strain is exceeded, it will break. Brittle fibers, such as graphite, have a low fracture strain and thus a limited ability to absorb energy. The following expression [16] calculates the energy required per unit area of the composite for tension fiber fracture:

$$u = \frac{V_f \sigma_{fu}^2 l}{6E_f} \tag{6.4}$$

V_f stands for fiber volume fraction, σ_{fu} stands for fiber ultimate strength, E_f stands for fiber modulus, and l stands for fiber length. For the energy-release rate produced by fiber breakage during the fracture process, Beaumont [17] presented a similar equation (with fiber length replaced by fiber critical length). This similarity is to be expected because the same elements that increase impact energy also increase fracture and durability of composites. Although fibers are essential for giving composites their high degree of strength, only a very small amount of the energy absorbed is due to fiber fracture. Experiments [17] have shown that the overall impact energy is not significantly affected by the number of fibers fragmented. It should be noted, too, that the presence of fibers significantly affects failure modes and, consequently, total impact energy.

6.4.3 MATRIX DEFORMATION AND CRACKING

The matrix around the fibers needs to break completely for the composite to fracture. Epoxies and polyesters, which are thermosetting resins, are fragile materials
that can only be slightly bent before breaking. Meta matrices, however, might
experience substantial plastic deformation. Although the matrix material can fracture and deform, both processes absorb energy; nonetheless, the energy needed
for plastic deformation is far greater than the contribution of surface energy. As a
result, metal matrices may make a large contribution to the overall impact energy
of composites, whereas polymer matrices may make a minor contribution.

The amount of work required to deform the matrix to rupture is related to the
amount of work required to deform the matrix to rupture per unit volume [18].
U_m times the matrix volume distorted per unit area of the crack surface. The
energy required for matrix fracture per unit area of composite is calculated using
Cooper and Kelly's [19] equation for the volume of matrix affected by fracture:

$$u = \frac{(1-V_f)^2}{V_f} \frac{\sigma_{mu}d}{4\tau} U_m \tag{6.5}$$

where σ_{mu} is the matrix's tensile strength, d is the diameter of the fiber, and τ is
the interfacial shear stress.

The contribution suggested by Eq. (6.5) to the overall impact energy in metal
matrix composites may be fairly large. However, because U_m is minimal in
brittle polymer matrices, the energy required for matrix deformation can be
generated via crack branching, in which case the cracks flow in the direction
opposite to the main direction of fracture. A matrix crack may branch to run
parallel to a strong fiber that is perpendicular to the fracture propagation direction or at a significant angle from it. Secondary cracks frequently form a surface
area that is significantly larger than the region parallel to major fissures. This
might potentially increase fracture energy by a factor of 10, which could be a
suitable strategy to increase composite toughness or the total amount of energy
absorbed during fracture.

6.4.4 FIBER DEBONDING

Cracks parallel to the fibers may detach the fibers from the matrix material during the fracture process (debonding cracks). During this process, the secondary
or chemical connections between the matrix material and the fibers are broken.
This type of cracking occurs when the contact is weak and the fibers are strong.
Depending on the relative strengths of the fiber and matrix, a debonding crack
may develop at the fiber-matrix interface or in the nearby matrix. In each case,
a new surface is produced. The fracture energy may be dramatically raised if
there is extensive debonding. A drop in contact strength may be accompanied
by an increase in impact energy because it encourages widespread debonding or
delamination [20].

Debonding fractures are secondary matrix cracks that originate from the primary cracks, as was already described. The figures Kelly [21] provided for the effort of debonding for various materials are typically s500 J/m^2 and on the order of the interface shear strength times the resin failure strain. The properties of the components and the interface can only be used to determine the debonding energy theoretically. According to Outwater and Murphy [22], Kelly [18] has also shown that the elastic energy retained in the fibers after debonding cannot be attributed to the energy of debonding.

6.4.5 EFFECT OF MATERIALS AND TESTING VARIABLES ON IMPACT PROPERTIES

Up until the middle of the 1960s, not much thought was given to studying how fiber composite materials behaved under impact. Early results on the topic were acquired without making any attempt to understand the phenomena of impact using a typical Charpy impact machine [23,24]. A fiber-reinforced polymer's impact behavior is predicted to be time-dependent, or reliant on the velocity of the hammer used to strike the specimen. Traditional impact-testing instruments lack the capability to study this key aspect of impact behavior. According to Rotem and Lifshitz [25], the ultimate strength of a glass-fiber-reinforced plastic increases with the rate of loading in the fiber direction. Later, using a specially made drop-weight impact-testing apparatus with easily adjustable experimental parameters, Broutman and colleagues [20,26–28] carried out extensive impact research. How to adjust and measure the various parameters listed in this section are explained in the discussion.

Broutman and colleagues looked into hybrid graphite-Kevlar-glass composites [20,26–28] as well as epoxy and polyester resins that are reinforced with glass fibers. Both systems employed layups that were cross-ply and unidirectional. In the studies, changes were made to the specimen dimensions, drop weight, impact velocity, fiber origination, and interface strength. High-speed photography was used to gain a significant understanding of the fracture process and related energy-absorbing devices.

The results on glass-fiber reinforced polymers and significant discoveries by other researchers are covered in detail in this subsection. The results and further details of hybrid composites are presented in chapter 7.

The impact behavior of composites is significantly impacted by the fiber orientation. Using E-glass-epoxy laminates, Mallick and Broutman [29] investigated how the angle of the fiber orientation affected the impact characteristics of off-axis composites. A [0/90/04/0]s unidirectional laminate has two layers that are placed underneath each surface layer and has the precise arrangement of [0/90/04/0]s, meaning it has two layers that are 90° to one another. The configuration of the cross-plied laminate is [(0/904/0)s]. Totally 13 plies, each 0.01 inch thick, are used in both systems.

Fibers in the outer layers of the laminates were cut at angles of 0°, 15°, 45°, 75°, and 90° with the longitudinal-beam axis to create rectangular examples. The fiber orientation angle is the name given to this angle (θ). The load was applied normally to the lamination plane in all cases, as indicated in Figure 6.6. In shattering the unidirectional specimens, the impact energy absorbed per unit width is illustrated in study [29].

At $\theta=60°$, the lowest value is found. The inclusion of 90° layers (beneath surface layers), in which fibers are oriented at tiny angles of $(90°-\theta)$ to the longitudinal-beam axis and thus capable of absorbing higher energy, increases the impact energy at $\theta=60°$. However, the Charpy impact test results on composites with 100% unidirectional fibers by Agarwal and Narang [30] show that the impact energy reduces with increasing fiber orientation. At $\theta=90°$, the minimum impact energy is measured. Study [29] shows the results of Mallick and Broutman [29] for cross-ply specimens. In this scenario, the absorbed energy curve is symmetric around $\theta=45°$, which is where the lowest impact energy is found.

Study [29] also depicts the impact energy of unidirectional specimens with the same drop height and specimen size. The impact energy absorbed by the cross-ply specimens is always more than that of the unidirectional specimens, with the exception of $\theta=0°$, where the unidirectional specimens absorb more energy. Lifshitz has established the impact strength of angle-ply composites [31].

Interface strength is another important material characteristic that has a big impact on composite failure mechanisms. Yeung and Broutman [20] modified the surface treatment of the glass fabric in order to modify the contact conditions. Polyester and epoxy resins were used to create the matrix material. The apparent shear strength from a short-beam shear-strength test was used to calculate the interface strength.

It was discovered that by adjusting the coupling agent on the fiber surface, the interface strength of polymer laminates may be adjusted over a wide range. The interface strength of epoxy laminates, on the other hand, could not be altered over the same range since the epoxy resin may form a strong bond with the glass surface even without a coupling agent. In an instrumented Charpy impact-testing equipment, polyester and epoxy laminates were tested to determine initiation energy (E_i), propagation energy (E_p), and total impact energy (E_t). The values of energies per unit area $[u_i=(E_i/bh), u_p=(E_p/bh),$ and $u_t=(E_t/bh)]$, where b and h represent specimen width and thickness, respectively, are plotted as a function of laminate shear strength in study [20].

It can be demonstrated that the initiation energy, u_i, rises with shear strength for both polyester and epoxy laminates. These fabric laminates exhibit enhanced interfacial bonding and higher intralaminar strengths, notably interlaminar tensile strengths, which is indicated by the rise in flexural strength as the shear strength rises. The epoxy laminates' stronger flexural strengths are reflected in their significantly higher initiation impact energy.

The propagation and total impact energy curves seem to be at their lowest points for polyester laminates. Over a threshold of interlaminar shear strength, the total impact energy increases with increasing shear strength. Impact energy drops as shear strength falls below a threshold level. Impact energy rises with increasing shear strength. Delamination was shown to be the primary mode of failure below the critical value of shear strength, while fiber failure appeared to be the dominant mode beyond the critical value, as illustrated in study [20]. The overall impact resistance of polyester laminates can be raised by decreasing the interfacial bonding. As a result, when the shear strength is at its weakest, the impact strength is at its highest. The considerable value of total impact energy is reached during the delamination phase, which takes place after failure commencement, when the interfacial connection is insufficient, it should be underlined.

The specimen can support significant deflections, which allows it to absorb more energy while supporting less load during propagation. Epoxy laminates do not have interfacial bonding that is sufficiently weak to result in significant delamination. The impact energy rises with shear strength over the critical shear-strength threshold, just like with polyester laminates. The highest impact energy recorded for an epoxy laminate with high shear strength is about equivalent to the impact energy recorded for a polyester laminate with low shear strength. In contrast to the low-shear-strength polyester laminate, which fails primarily due to delamination, the epoxy laminate generally fails due to fiber failure.

Yeung and Broutman [20] point out that the interlaminar fracture surface work is not much affected by the surface treatment. As a result, more extensive delamination rather than a change in the delamination process is what causes the increase in impact energy caused by a reduction in contact strength.

The strong energy-absorbing properties of laminates with weak interfacial bonding are of particular importance when fiber-reinforced polymers are used as armor materials. In some applications, high-impact resistance (total energy absorbed) is essential. In some applications, a low contact strength can be especially advantageous due to the rise in total energy absorbed.

6.5   ENVIRONMENTAL-INTERACTION EFFECTS

The material behavior under cyclic and impact loads was covered in the two sections before this one. It's also critical to understand how materials react under various conditions, such as in corrosive environments, at low and high temperatures, and in terms of their long-term physical and chemical stability. This section focuses on how different environmental factors cause composite materials to degrade.

Numerous factors, such as stress corrosion of the reinforcing fibers, degradation of the fiber-matrix interface, and loss of adhesion and interfacial bond strength, can contribute to the degradation of composite materials. (1) Chemical deterioration of the matrix material, (2) Temperature and time dependence of matrix strength, and (3) Rapid deterioration as a result of temperature and chemical environment.

These factors lead to the conclusion that composite materials are no longer useful when the stiffness is reduced to the point where structural instability and/ or failure or rupture occur. Environmental factors have an effect on the matrix material, the interface, and the fibers all at once. As a result, composites age not only as their constituent components do but also as the way they interact with one another. The impacts of environmental factors on fibers, matrix material, and interface are covered in the subsections that follow.

6.5.1 FEATURES OF STRESS CORROSION

The strength of materials is theoretically constrained by the quantity of forces holding atoms together. However, most solids' strengths are substantially weaker in actuality than they are in theory. The discrepancy between theoretical and experimental strengths is attributed to errors or weaknesses in the material. In addition, under the right circumstances, many hard amorphous or crystalline solids show delayed failure, where the strength is greatly affected by the duration of the load application. For instance, silicate glasses have a decreased breaking strength in humid settings, and many polymers lose strength when exposed to solvents and washing solutions. The majority of studies have found that delayed failure under continual load results from defects growing to a critical magnitude under the influence of a reactive environment, where the stress state at the most significant defect is sufficient to cause a spontaneous failure.

The most frequent flaws in glasses and other amorphous materials are surface fractures or other faults that might develop as a result of tension and chemical damage. On the other hand, the degree of stress and the bulk qualities of the material define how significant a fault is. A noncritical fault may develop into a critical one due to flaw development, higher stress, or alterations in bulk properties over time.

In general, little is known about the mechanism of stress corrosion for different materials and circumstances. But all systems have a few things in common:

1. In inert settings or at low temperatures, when corrosion reaction rates are expected to be negligibly slow, material breaking strengths become independent of load application duration and constantly maintain a relatively high value. At high loading rates that do not provide enough time for the corrosion process to occur, equivalently high strengths are seen.
2. Exposing these materials to reactive conditions before, but not during, a test had a less impact on the test findings, implying that stress accelerates corrosion.
3. When these materials are exposed to reactive conditions during a test, they experience delayed rupture at stresses that are significantly lower than features 1 and 2.
4. The relationship between the time of loading and the failure stress is continuously influenced by temperature. In general, as the temperature rises, the strength of the material decreases.

6.5.2 STATIC FATIGUE AND STRESS-RUPTURE OF FIBERS

The most popular fiber-reinforcing substance is glass. When subjected to static loads, glass is known to fail gradually. Prior to failure, there are no indicators of creep at room temperature, but the majority of researchers think that moisture accelerates the spread of preexisting defects under prolonged stress. A defect that reaches a critical size immediately fails. There is a ton of knowledge available about glass and glass fibers. Glass fibers are the subject of a review essay by Lowrie [32]. E-glass fibers were subjected to static fatigue tests by Otto [33] at both normal and high temperatures.

His results are shown in study [20]. Although the data are highly variable, particularly at lower temperatures, accurate predictions of the fatigue rate can be made. The strength decreased by 40,000–65,000 psi during the course of the studies, which ranged in duration from 1 minute to 20 hours (280–435 MPa). The glass fibers become weaker as the temperature rises. A sizeable amount of their short-term strength is lost at higher temperatures.

The fibers at room temperature lose around 3% of their short-term strength for each tenfold increase in load application time. Applications for reinforced composite materials that call for high modulus and strength even at high temperatures use boron and graphite filaments. An early review of boron filaments was written by Wawner [34]. Boron filaments' stress-rupture characteristics haven't gotten a lot of attention. A few stress-rupture test results on single boron filaments at 900°F were published by Cook and Sakurai [35] (Study [35]). As the load time increased from 3 to 5 hours, the boron filaments showed a considerable reduction in stress. The trials were conducted in both an air and an argon atmosphere.

The boron filaments performed better when tested in an inert atmosphere at the same stress intensity and temperature. The stress level significantly decreased after a load period of 10–20 hours, as seen in study [35]. The data in study [35] seem to suggest that even at short load durations in air, a significant loss in strength occurs. By contrast, a tensile test (short-term strength) up to 900°F demonstrated almost no difference between the air and argon atmospheres. For stress-rupture tests in air lasting longer than 20 hours, there are no data available. According to the findings in study [35], strength degradation occurs significantly under a constant load at 900°F. In an argon atmosphere, filaments lose approximately 75% of their initial strength in less than 100 hours, whereas in an air atmosphere, they lose the same amount in only 10 hours.

The utilization of brittle high-strength fiber materials is constrained by a time-dependent distribution of fault sizes. If a fibrous composite solely experiences fiber strength degradation, the longitudinal strength of a unidirectional composite will be significantly compromised but the transverse and shear strengths will only be marginally affected. The advantages of these high-strength reinforcing materials can be effectively utilized by coating fibers or inserting them into a matrix material.

The coating, or "coupling agent," shields the fibers from abrasion and other sources of surface imperfections during manufacturing. Along with the matrix material, the coupling agent serves as a barrier between the harsh environment and the reinforcement. A chemical composition that is inert to the anticipated service environment is another goal when designing high-strength reinforcements.

6.6 APPLICABLE PROBLEMS

1. Describe the progression of fatigue damage in the following laminates: $[0]_8$ and $[0/\pm45/90]$. Assume cyclic stress in tension ($R=0.1$) equivalent to 75% of the laminate ultimate strength.

2. Fatigue strengths with zero mean stress of a glass-polyester composite at 10^3, 10^4, 10^5, and 10^6 cycles have been found to be 84, 70, 60, and 52 MPa, respectively. Stress-rupture strength of the same composite follows the relationship:

$$Sc = 21.8 - 4\log t$$

where Sc is the stress rupture strength in megapascals and t is the period in minutes. Create a master diagram [Eq. (6.1)]r using these data and the Goodman-Boller relationship to show the impact of mean stress on permitted stress amplitude at various cycle lifetimes. Assume that 2,000 cycles per minute were used for the fatigue tests. If the laminate whose fatigue characteristics are shown in Figure 6.6 is interrupted after 30,000 cycles and subjected to a tensile strain of 2%, calculate whether it will fracture at this strain level.

3. Describe why a similar glass-fiber composite fails with significant delamination while the high-modulus unidirectional graphite-fiber-reinforced epoxy beam fractures cleanly into two halves without delamination and with little fiber pullout.

REFERENCES

[1] J. W. Dally, and B. D. Agarwal, "Low cycle fatigue behavior of glass fiber reinforced plastics," *Proceedings of the Army Symposium on Solid Mechanics,* AMMRC MS 70–75, New York. 1970.

[2] L. J. Broutman, and S. Sahu, "Progressive damage of a glass reinforced plastic during fatigue," *SPI, 24th Annual Technical Conference,* Washington, DC, 1969, Sec. 11-D.

[3] M. J. Owen, and T. R. Smith, "Some fatigue properties of chopped-strand-Ma\polyester-resin laminates," *Plast. Polym.,* 36, 33–44 (1968).

[4] T. Tanimoto, and S. Amijima, "Fatigue properties of laminated glass fiber composite materials," *SPI, 29th Annual Technical Conference,* Washington, DC, 1974, Sec. 17-B.

[5] T. Tanimoto, and S. Amijima, "Progressive nature of fatigue damage of glass fiber reinforced plastics," *J. Compos. Mater.,* 9(4), 380–390 (1975).

[6] H. T. Hahn, and R. Y. Kim, "Fatigue behavior of composite laminate," *J. Compos. Mater.,* 10(2), 156–180 (1976).

[7] J. W. Dally, and L. J. Broutman, "Frequency effects on the fatigue of glass reinforced plastics," *J. Compos. Mater.,* 1, 424–442 (1967).

[8] L. Cessna, J. Levens, and J. Thompson, "Flexural fatigue of glass reinforced thermoplastics," *SPI, 24th Annual Technical Conference,* Washington, DC, 1969, Sec. 1-C.

[9] L. J. Broutman, and S. Sahu, "A new theory to predict cumulative fatigue damage in fiberglass reinforced plastics," *Composite Materials: Testing and Design (Second Conference)*, ASTM STP 497, American Society for Testing and Materials, Philadelphia, PA, 1972, pp. 170–188.

[10] J. F. Mandell, and U. Meier, "Fatigue crack propagation in 0°/90° E-glass/epoxy composites," *Fatigue of Composite Materials*, ASTM STP 569, American Society for Testing and Materials, Series: Report (Massachusetts Institute of Technology. Sea Grant Project Office); no.MITSG 73-14, Philadelphia, PA, 1975, pp. 28–44.

[11] J. F. Mandell, "Fatigue crack propagation in woven and non-woven fiberglass laminates," *Composite Reliability*, ASTM STP 580, American Society for Testing and Materials, Philadelphia, PA, 1975, p. 515.

[12] Z. Hashin, and A. Rotem, "A fatigue failure criterion for fiber reinforced materials," *J. Compos. Mater.*, 7(4), 448–464 (1973).

[13] B. D. Agarwal, and J. W. Dally, "Prediction of low-cycle fatigue behaviour of GFRP: An experimental approach," *J. Mater. Sci.*, 10(1), 193–199 (1975).

[14] S. S. Manson, "Fatigue: A complex subject-some simple approximations," *Exp. Mech.*, 5(7), 193–226 (1965).

[15] J. F. Tavemelli, and L. F. Coffin, Jr., "Experimental support for generalized equation predicting low cycle fatigue," *J. Basic Eng.*, 84(4), 533–541 (1962).

[16] B. D. Agarwal, L. J. Broutman, and K. Chandrashekhara, *Analysis and Performance of Fiber Composites*, Wiley, 2017.

[17] P. W. R. Beaumont, "A fracture mechanics approach to failure in fibrous composites," *J. Adhes.*, 6, 107–137 (1974).

[18] A. Kelly, *Strong Solids*, Clarendon, Oxford, 1973.

[19] G. A. Cooper, and A. Kelly, "Tensile properties of fibre-reinforced metals: Fracture mechanics," *J. Mech. Phys. Solids*, 15, 279 (1967).

[20] P. Yeung, and L. J. Broutman, "The effect of glass-resin interface strength on the impact strength of fiber reinforced plastics," *Polym. Eng. Sci.*, 18(2), 62–72 (1978).

[21] A. Kelly, "Interface effects and the work of fracture of a fibrous composite," *Proc. R. Soc.*, A319, 95 (1970).

[22] J. O. Outwater, and M. C. Murphy, *24th Annual Technical Conference of the Society of Plastics Industry*, Washington, DC, 1969.

[23] A. J. Barker, "Charpy notched impact strength of carbon-fiber/epoxy-resin composites," *First International Conference on Carbon Fibers*, London, 1971, Paper 20.

[24] G. R. Sidney, and F. J. Bradshaw, "Some investigations on carbon-fibre reinforced plastics under impact loading and measurement of fracture energies," *First International Conference on Carbon Fibers*, London, 1971.

[25] A. Rotem, and J. M. Lifshitz, "Longitudinal strength of unidirectional fibrous composite under high rate of loading," *SPI, 26th Annual Technical Conference*, Washington, DC, 1971, Sec. 10-G.

[26] L. J. Broutman, and A. Rotem, "Impact strength and toughness of fiber composite material," *Foreign Object Impact Damage to Composites*, ASTM STP 568, American Society for Testing and Materials, Philadelphia, PA, 1975, pp. 114–133.

[27] L. J. Broutman, and A. Rotem, "Impact strength and fracture of carbon fiber composite beams," *SIP, 28th Annual Technical Conference*, Washington, DC, 1973, Sec. 17-B.

[28] L. J. Broutman, and P. K. Mallick, "Impact behavior of hybrid composites," AFOSR TR-75-0472, November 1974.

[29] P. K. Mallick, and L. J. Broutman, "Impact properties of laminated angle ply composites," *SPI, 30th Annual Technical Conference*, Washington, DC, 1975, Sec. 9-C.

[30] B. D. Agarwal, and J. N. Narang, "Strength and failure mechanism of anisotropic composites," *Fiber Sci. Technol.*, 10(1), 37–52 (1977).

[31] J. M. Lifshitz, "Impact strength of angle ply fiber reinforced materials," *J. Compos. Mater.*, 10(1), 92–101 (1976).

[32] R. E. Lowrie, "Glass fibers for high strength composites," in L. J. Broutmai, and R. H. Krock, eds., *Modem Composite Materials*, Addison-Wesley, Reading, MA, 1967.

[33] W. H. Otto, "Properties of glass fibers at elevated temperatures," *Proceedings of 6th Sagamore Ord. Materials Research Conference*, Washington, DC, 1959, p. 277.

[34] F. W. Wawner, Jr., "Boron filaments," in L. J. Broutman, and R. H. Krock, eds., *Modem Composite Materials*, Addison-Wesley, Reading, MA, 1967.

[35] J. L. Cook, and T. T. Sakurai, *SAMPE National Symposium, 10th*, 1966, 10, Sec H-1.

7 Discontinuous Basalt Fiber-Reinforced Hybrid Composites

7.1 BASALT FIBERS

Basalt fiber is a relative newcomer in the world of fiber-reinforced polymers (FRPs) and structural composites. Its chemical composition is similar to glass fibers, but it has stronger characteristics and is much more resistant to attack from alkaline, acidic, and salt solutions than most glass fibers, making it a good option for concrete, bridge, and beach buildings.

Lava at a planet's surface rapidly cools, becoming basalt, a type of volcanic rock. The Earth's crust contains the most of this type of rock. The lava source, rate of cooling, and previous exposure to the elements all affect the basalt rock's characteristics. High-quality fibers are made from basalt deposits with stable chemical compositions.

The manufacturing of glass fiber is similar to that of basalt. Crushed basalt rock is the only basic material required to manufacture the fiber. At a temperature of about 2,700°F, volcanic basalt rock is melted to produce a continuous thread.

7.1.1 Characteristics, Applications

The acidity modulus, M, of rock wool fibers, which expresses the ratio of acidic to basic oxides, can be used to identify these fibers. If $M_s < 1.2$, the fiber is referred to as slag wool, and its primary component is cinder. Because they are highly fragile and exhibit poor chemical resistance, such subpar fibers are no longer made nowadays. The fiber is thought to be mineral wool if $M_s = 1.2$–1.5, and its primary constituents are cinder and basic volcanic rock. Although brittle, these fibers have good insulating qualities, and therefore, their relevance in the construction industry is substantial. The fiber is known as rock wool if $M_s > 1.5$, and basalt wool if its foundation material is basalt (basalt fiber, BF). Basalt, a volcanic, effusive, over-ground rock that is saturated with 45–52 wt% SiO_2, forms the basis of basalt fiber.

Basalt has a number of fantastic qualities because of how it was formed. Its fibers have a large capacity for heat and sound dampening and excel as vibration isolators in addition to having a high elasticity modulus and great heat resistance.

Although BF has been widely used as an insulating material for nearly 50 years, there are significantly fewer studies about it in the literature than about GF and

DOI: 10.1201/9781003429197-7

CF. Extensive reports on GF and CF addressed their manufacture, characteristics, and applications.

However, very little research has been done on BF, and even less is known about their incorporation into polymer matrix. In the late 1970s and early 1980s, scientists from the former Soviet Union published the first publications.

The authors looked at BFs as insulation and as textile industry raw materials [1,2]. It's interesting to note that the insulating layer of the Soviet Union's astronauts' spacesuits at the time was made of basalt fibers.

It should be emphasized that before discussing the state of the art with BF and BF applications in composites, all articles used high-quality continuous BFs or their chopped counterparts. Subramanian and Austin [3] made the initial mention of BF as a viable polymer reinforcing material when they asserted that BF can replace GF in polymer matrix composites. The authors examined the BF-to-matrix adhesion in single BF-reinforced thermosetting polyester matrix composites since they recognized its significance.

It was discovered that silane surface treatment of BF improved adherence to the matrix, which was evident in the bending strength of the composites. After embedding a single BF that had been treated with different forms of silane in an epoxy resin, Park and Subramaniam noticed that 3-aminopropyl-triethoxy-silane and dimethoxysilane increased the interfacial shear strength [4,5].

Questions like, "What are the useful features and the limitations of BFs, what are the application domains, and what are their advantages over other fibers," became more and more common as BFs were tested as reinforcing materials in the 1990s. When considering the service life of thermal insulation [6], Wojnarovits compared the corrosion resistance of BF and GF, noting that resistance to water, pollution, and high temperatures is essential.

It was discovered that if the design is inadequate, corrosion will severely damage both BF and GF, causing the insulating material to fail. Additionally, it was discovered that the fibers' chemical resistance is unrelated to their mechanical characteristics. The parameters affecting the mechanical characteristics of silicate fibers such as glass, ceramic, kaolin, and basalt were clarified in this study. The best mechanical performance was displayed by the fibers with the smallest pore volume and, in particular, the smallest ratio of big mesopores (about 0.03–0.15 pm). The chemical makeup of the fibers and their mechanical qualities, on the other hand, did not directly correlate. According to Jung and Subramanian's investigation into the tensile strength of BF, adding alumina (2 wt%) to basalt increases the fiber strength by ca. 15% [7].

The subject of whether basalt is damaging to health was brought up by the expanding use of basalt, particularly as an insulating material in the construction and automobile industries. Kogan and Nikitina [8] conducted a comparison of basalt fiber and asbestos. For 6 months, they made rats breathe air laced with asbestos and basalt fibers. A dose of 1.7 g/kg of asbestos fibers caused one-third of the rats to perish, whereas a dose of 2.7 g/kg resulted in the death of every rat. In the instance of BF, the animals continued to live even after receiving a dose of

10 g/kg. Similar studies were carried out by Adamis and McConnell [9,10]. They also concluded that BFs do not present any risk to people.

The majority of the articles on BF-reinforced composites were published in the second half of the 1990s by Park et al. from Korea. They used the techniques of single fiber fragmentation and acoustic emission to assess polymer composites made from polycarbonate (PC) [11,12] and epoxy resin (EP) [13,14]. The number of acoustic events and the interfacial shear strength value obtained from the fragmentation tests were shown to have a strong correlation in the case of PC matrix composite (where aminosilane was utilized as a treatment agent).

Polypropylene (PP) matrix/short BF composites were made by Bashtannik et al. [15] using an extrusion technique. The same authors came to the conclusion that composites with BF reinforcement have excellent wear characteristics. When no coupling agent was applied, Botev et al. [16,17] found that the mechanical characteristics of PP matrix/short BF composites were extremely subpar.

The main paper written by Goldsworthy and published in the August 2000 issue of the *Journal of Composites Technology* provided more motivation to create BF-reinforced composites [18]. The title of this essay on the origins, development, and prospects of polymer composites was "New Basalt Fiber Increases Composite Potential." The author projected the widespread use of BF and highlighted its potential as a new natural-source reinforcing material.

Russian basalt's chemical composition was studied by Gurev and Morozov [19,20] in relation to the manufacturing of continuous BF. The explanation was that the mechanical properties of BF were significantly influenced by technological parameters. The basalt rock with the highest SiO and Al_2O_3 content and the lowest CaO and MgO content was found to be the most appropriate for continuous BF manufacture.

Short (1–3 mm) BF-reinforced polystyrene (PS) composite plates were created by Zihlif and Ragosta [21], who discovered that while the impact resistance and Young's modulus grew monotonically with increasing BF content, the strength of the composite reached a maximum. In BF composites with a high-density polyethylene (HDPE) matrix, Bashtannik et al. [22] looked into the interfacial adhesion between the fiber and matrix. It was discovered that, compared to conventional fiber-reinforced systems, the characteristics of BF-reinforced polymer composites were significantly more sensitive to the BF surface treatment.

On the basis of the aforementioned literature review, it can be stated that BF might replace asbestos fibers that are already illegal if good adherence to the polymer matrix can be supplied in addition to adequate fiber quality and affordability.

7.1.2 BASALT FIBERS MADE BY MELT-BLOWN PROCESS AND THEIR CHARACTERISTICS

The BF utilized in the thermoplastic and thermoset composites discussed here was produced through the so-called Junkers process (Toplan Ltd, Tapolca, Hungary). The key part of this strategy is feeding basalt melt from a 1,580°C furnace to a horizontal shaft fiber spinning machine. This consists of three centrifugal heads,

TABLE 7.1

Properties of the Reinforcing Fibers

Fiber	(g / cm³)	D_{av} (µm)	σt (MPa)	ε_t (%)	E_t (GPa)
Basalt	2.70	9.0 ± 2.7	586 ± 267	1.12 ± 0.45	60.4 ± 18.9
Glass	2.54	12.2 ± 1.4	1,540 ± 556	3.10 ± 1.18	52.2 ± 18.2
Carbon	1.76	7.9 ± 0.9	2,372 ± 977	0.02 ± 0.01	157.9 ± 58.6
Ceramic	2.55	5.9 ± 1.1	828 ± 306	1.50 ± 0.62	59.9 ± 13.0

one accelerating cylinder, and two fibrillizing cylinders. High-pressure air blasts the centrifugal force-produced fibers off, as shown in study [23].

This fiber-spinning process is incredibly efficient and reasonably priced. However, it has the disadvantage that, when the fibers gradually cool, smaller or larger "heads" remain at their ends, which has an adverse effect on the strength and toughness. Scanning electron microscopy (SEM) images acquired from the fracture surface of a BF-reinforced PP composite show the impact of the BF head (shown in study [23]). To avoid any negative effects, broken heads were removed by settling in water. This was a crucial step, especially in light of the performance of the hybrid reinforcement. Single fibers were used to assess the tensile strength and Young's modulus of BF (Toplan Ltd., Tapolca, Hungary) in accordance with the JIS R 7601 standard. The melt-blown discontinuous BF's properties, as well as details on GF, CF, and CeF, are listed in Table 7.1. In addition to maintaining the fibers' clamping length at 20 rnm, the initial surface polish was also removed for better comparison (by burning in a furnace at 400°C for 4 hours).

7.2 HYBRID COMPOSITES

The word "hybrid," which has Greek and Latin roots, is used in a wide range of scientific disciplines. However, this phrase is most frequently used in biology to refer to the hybridization of various animal and plant species. Similar to this, the engineering sciences use this word in a variety of contexts, but the general idea is that this term refers to a "mixing" of several elements. Hybrid composites, in the context of polymer composites, are those systems in which one type of reinforcing material is incorporated in a combination of different matrices, or in which two or more reinforcing and filling materials are present in a single matrix, or in which both approaches are combined [24–26].

It has long been a focus of research to customize the mechanical properties of composites by using various reinforcing fibers. Be aware that the properties of the resulting composites may be positively or negatively impacted by the hybrid idea. Up until the middle of the 1990s, the research mostly concentrated on GF, CF, and AF fiber (reinforcement) hybridization. Early papers by Bunsell and Harris [27], Summerscales and Short [28], and others showed that the hybrid effect is most pronounced in composites containing both GF and CF. It was determined that

GF can significantly lower the cost of CF-reinforced composites while increasing impact resistance. On the other hand, when CF is added to composites that contain GF, the flexural modulus significantly rises. The impact resistance and flexural strength of CF-reinforced epoxy matrix (EP) composites were increased by Marom et al. [29,30] by adding AF. Chaudhuri and Garala [31] discovered that by just adding 15 wt% GF, the compressive strength of CFIEP composites could be optimized. The research done by Fu et al. [32] on PP matrix-based hybrid fiber composites provided support for this conclusion. They looked at how the GFICF ratio affected the Charpy impact energy of injection-molded specimens reinforced with 25 wt% short fibers. It was discovered that the GFICF ratio in the hybrid composites affects the mean fiber length. The fibers were longer if the relative GF content was higher, but higher relative CF concentrations caused the fiber length to decrease because of melt processing (extrusion, injection molding).

It should be noted that brittle fibers break during processing due to interactions between fibers and between fibers and machine parts (screws, barrels). In the Charpy experiments, there was a little increase in impact energy as the GF content rose.

Jang and Lee [33] demonstrated the favorable impact of CF on the flexural strength of polyphenylene sulfide (PPS)/GF composites. They looked at how the GFICF ratio affected the flexural strength and modulus of isotropic specimens. It was discovered that flexural strength and modulus were directly correlated with the CF concentration.

Sohn and Hu [34] looked at the possible modes of failure for continuous CFEP composites.

When AF (15 wt%) was introduced, the delamination strength increased by more than 100% while the compressive strength declined. Hybrid fiber composites don't just benefit from improved mechanical qualities. Friedrich and Jacobs [35] investigated the steel-facing friction qualities of EP, polyetheretherketone (PEEK), and polyamide (PA) matrix composites supplemented with GF, CF, and AF fibers. The hybrid composites made of AFICF had the best wear resistance. Only in the second half of the 1990s did reports of hybrid composites with reinforcements other than the conventional fibers surface in the literature (GF, CF, AF). The reason for this is that natural fibers have become more popular as a result of environmental concerns. Sisal/GF hybrid fiber LDPE matrix composites were investigated by Kalaprasad et al. [36,37] at various fiber orientations. The impact characteristics of flax/glass/PP matrix hybrid composites made by hot compression molding from a needle-felted preform were studied by Benevolenski [38].

The matrix and reinforcement were both present in this felt in the form of short fibers. It was discovered that even a little amount of GF (45 wt% flax, 5 wt% glass) significantly improved the dynamic impact characteristics. A twofold increase in impact resistance was seen when flax and GF content were combined at 20 and 30 wt%, respectively.

A few more recent studies, mostly based on the work of Asian experts, deal with natural fibers of tropical origin and glass fiber-reinforced hybrid composites.

Rozman et al. [39] looked at PP composites reinforced with GF and oil palm fruit bast. According to a report, in order to improve the mechanical properties of the hybrid composites, PP must be maleated and natural fibers must have their surfaces treated with trimethoxysilyl-propylmethacrylate.

Rout et al. [40] studied hybrid composites made of coir, GF, and thermosetting polyester (13/7/80 wt%, respectively). The hybrid composites' tensile strength may be significantly increased, and the moisture uptake could be drastically reduced by alkaline-treating coir. In the cases of bamboo fiber/GF/PP and pineapple/sisal/GF/polyester hybrid composites, Thwe and Liao as well as Mishra [41,42] observed the same trend (reduction in moisture uptake and modest improvement in mechanical characteristics).

7.3 PROPERTY PREDICTION

Early theoretical analyses of the hybrid effect revealed that calculations based on the straightforward mixing rule conflicted with the experimental findings, indicating the intricacy of the "hybridization." In order to describe the mechanical characteristics of hybrid composites, the rule of hybrid mixtures (Eq. 7.1):

$$P_H = P_1V_1 + P_1V_2 \tag{7.1}$$

where P_H is any mechanical parameter of the hybrid composite, $P_{1,2}$ and $V_{1,2}$ are the qualities of the 1 and 2 simple composites, respectively, and $V_1 + V_2$ is the volume fraction of the 1 and 2 reinforcing fibers. Because the strength of the fiber-matrix interfacial adhesion and the interaction of the fibers that make up the hybrid composite cannot be calculated using the simple rule of mixtures when it comes to hybrid composites, experimental results should be considered.

The majority of research in this area focuses on how the so-called hybrid effect manifests and describes how the reinforcing materials work together synergistically. Marom et al. [43] claim that a positive hybrid effect takes place when a composite's attributes are superior than the values calculated using the rule of mixtures, whereas a negative hybrid impact shows up when the calculation using the rule of mixtures yields a higher value than the actual one. Numerous researchers measured the positive and negative impacts of hybrid composites with various reinforcing and matrix materials, but Fu's [44,45] findings are the most noteworthy.

7.4 APPLICATIONS

Hybrid composites are used extensively because of their beneficial qualities and the wealth of literature on the subject. For pipes and tanks, Zhu et al. [46] substituted epoxy matrix hybrid composites reinforced with GF/CF for steel. In addition to being corrosion-resistant and strong enough, these composite parts showed good creep resistance. Due to its outstanding energy absorption capacity, Chiu et al. [47] advised the use of braided AFICF-reinforced epoxy matrix hybrid composite tubes and profiles in the automobile industry. Since their research revealed that the fatigue strength of these materials in water is higher than that of the composite containing only GF, Shan and Liao [48] advised using GF/CF/epoxy hybrid composites in marine applications. The development of the LAK series of glider planes in 2000 marked the beginning of a new era in the history of sporting plane manufacture. This plane can fly at a top speed of 270 km/h and weighs just 200 kg when it is constructed from GF/CF/AF-reinforced panels. The newest use for hybrid composites is in the wheel blades of wind power plants [49]. A 9 m-long hybrid composite wind blade with GFICF reinforcement was created in the USA in 2003. Hybrid composites are anticipated to be produced on a big scale, particularly as parts with huge dimensions.

7.5 THERMOPLASTIC HYBRID COMPOSITES

By melt compounding in a Brabender internal mixer, different short (nominal beginning length of 20 mm) fiber PP composites were created. Remember that Table 7.1 lists the employed fibers along with their properties. Table 7.2 provides a summary of the prepared compositions.

The fibers were added to the PP melt ($T = 190°C$, kneader revolution 50 rpm, mixing duration = 3 minutes) and homogenized for an additional 5 minutes to prevent significant fiber breaking. Compression molding was used to create composite test sheets from the compounds. Using a 3-point bending configuration, static (subscript "s") and dynamic (subscript "d") flexural tests were performed on specimens cut from the sheets. Static testing was conducted with a 1 mm/min deformation rate.

Additionally, linear elastic fracture mechanics was used to calculate the static and dynamic fracture toughness (K, and K, respectively) values, and notched specimens with different notch lengths were employed. Table 7.3 provides an overview of the flexural strength, modulus, E, and K values.

The CF-reinforced PP composites demonstrated the best attributes as expected because CF has the highest strength and modulus among all investigated fibers (Table 7.1). However, if the production costs are also taken into account, it should be noted that these features come at a considerable cost because CF is 1.5 orders of

TABLE 7.2
Reinforcement Content and Composition of the Hybrid Composites

					Batch					
Fiber	1	2	3	4	5	6	7	8	9	10
PP	100	70	70	70	70	70	70	70	70	70
BF	-	30	-	-	20	20	20	10	10	10
GF	-	-	30	-	-	10	-	-	-	-
CF	-	-	-	30	-	-	10	-	20	-
CeF	-	-	-	-	30	-	-	10	-	20

TABLE 7.3

The Hybrid Fiber-Reinforced Composites' Mechanical Characteristics

Batch	σ_s (MPa)	σ_d (MPa)	E_s (GPa)	E_d (GPa)	K_s (MPam$^{0.5}$)	K_d (MPam$^{0.5}$)
1	36.6 ± 1.4	26.6 ± 2.3	1.1 ± 0.0	0.9 ± 0.1	2.2 ± 0.3	4.6 ± 1.0
2	26.2 ± 0.5	23.0 ± 0.8	2.4 ± 0.1	1.0 ± 0.1	1.1 ± 0.1	4.0 ± 0.6
3	40.4 ± 3.5	38.1 ± 6.5	2.6 ± 0.2	1.2 ± 0.1	1.9 ± 0.3	6.0 ± 0.6
4	58.3 ± 4.5	51.3 ± 7.4	4.1 ± 0.4	1.9 ± 0.1	2.3 ± 0.1	7.3 ± 0.8
5	43.6 ± 2.2	31.0 ± 1.5	2.5 ± 0.2	1.1 ± 0.1	2.0 ± 0.1	4.3 ± 0.6
6	31.9 ± 3.5	25.1 ± 3.4	2.3 ± 0.2	1.3 ± 0.2	1.6 ± 0.1	3.1 ± 0.5
7	32.2 ± 3.3	31.7 ± 2.1	2.7 ± 0.1	1.5 ± 0.1	1.7 ± 0.1	4.9 ± 0.4
8	31.0 ± 2.0	22.1 ± 0.7	2.7 ± 0.0	0.9 ± 0.0	1.3 ± 0.1	3.2 ± 0.8
9	31.7 ± 4.2	29.6 ± 3.2	2.5 ± 0.4	1.3 ± 0.2	1.5 ± 0.2	3.8 ± 0.4
10	43.3 ± 5.6	40.8 ± 2.5	4.0 ± 0.1	1.8 ± 0.2	2.2 ± 0.2	6.6 ± 0.7
11	26.7 ± 2.0	24.9 ± 2.0	2.4 ± 0.1	0.8 ± 0.1	1.3 ± 0.2	2.4 ± 0.3

magnitude more expensive than BF. The practical application of BF in hybrid composites and the significant cost reduction of the materials are demonstrated by the excellent properties of the composite containing 20 wt% CF and 10 wt% BF. This assertion is supported by the RHM, which demonstrates that a favorable hybrid effect was only obtained in the composite having 20 wt% CF and 10 wt% BF.

Surprisingly, the CeF-containing and CeF-reinforced composites had the worst characteristics. This result is probably a result of significant CeF breakdown during compounding. Keep in mind that CeF costs approximately the same as CF. The results also showed that all composites had enhanced stiffness, and the hybrid composites reinforced with BF/CF had increased strength.

SEM micrographs of the fracture surfaces show that fiber pull-out is the most common failure mode for short-fiber hybrid composites. Remember that the strength of the interfacial adhesion has a significant impact on fiber pull-out. The fibers' greatest pull-out length was approximately 100 pm, and the voids that remained after the fibers were pulled out are indicative of poor interfacial adhesion.

Since they have essentially identical strength qualities and densities, it was indicated at the beginning of this chapter that inexpensive BF can be used as a substitute for GF (the price of BF is around one-third that of GF). However, if it is not properly finished (surface coated, etc.), BF is a less effective reinforcement than GF. On the other hand, consideration needs to be given to the fact that the melt-blown BF employed is a lesser grade.

7.6 THERMOSET HYBRID COMPOSITES

BF is probably more suited for thermoset reinforcement than for thermoplastics because of its inherently brittle nature. The mechanical loading of the reinforcement is noticeably lower during resin infiltration than during melt compounding techniques. Basalt fiber mat-reinforced thermoset composites were made using resin transfer molding to test this idea (RTM). Keep in mind that the reinforcing mat was usually a felt made of short strands, with needling used to guarantee proper handling.

7.7 BASALT FIBER MAT-REINFORCED HYBRID THERMOSETS

Recently, interpenetrating network (IPN) structure hybrid thermosets were created. Although the main objective was to increase toughness through IPN production, the associated resins were crucial for composite applications. Vinylester (VE) and amine-crosslinkable EP resins work well together because the resulting thermosets have a low viscosity, cure quickly, and have an IPN structure that is advantageous for composites. Note that the inclusion of styrene, the crosslinking agent for VE, results in low viscosity, which aids in wetting out the reinforcement. The production cycle time can be shortened due to VE's quick curing as compared to EP alone.

However, the IPN structuring provides the most fascinating component. The matrix structure can be used to achieve an intermittent bonding of the reinforcing fiber due to the average width of the IPN forming bands being around 100 nm. Intermittent indicates that if the fiber surface is properly "processed" for one (i.e., VE or EP) of the resin components, good-poor-good bonding may take place on the nanoscale. RTM created BF mat-reinforced composite sheets to test the validity of this idea. The BF mat with discontinuous fibers (length 2.50 mm) has a different surface chemistry.

The original sizing (finish) of the BF was removed by burning in order to change the sizing (finish) (500°C, 3 hours). The BF's organ silanes were sized in aqueous solutions with silanes that had either vinyl (VS) or epoxy (ES) functionalities. Rectangular specimens were cut from composite sheets that contained around 30 wt% BF mat with a variety of surface treatments, and they were then loaded under standardized tension and flexural conditions. Table 7.4 provides a summary of the relevant findings.

TABLE 7.4
Mechanical Characteristics of IPN-Structured BF Mat-Reinforced VE/EP Resins

Component	Loading	σ (MPa)	E (GPa)	ε (%)
VE/EP matrix	T	40.8 ± 1.7	2.4 ± 0.1	4.5 ± 0.8
	F	75.9 ± 1.5	2.2 ± 0.1	5.2 ± 0.2
With as-received BF mat	T	47.5 ± 3.8	4.8 ± 0.6	2.1 ± 0.4
	F	100.3 ± 3.1	6.2 ± 0.2	2.1 ± 0.1
With burned BF mat	T	44.3 ± 0.9	4.3 ± 1.0	0.9 ± 0.2
	F	77.2 ± 9.1	5.8 ± 0.9	3.0 ± 0.7
With ES-treated BF mat	T	58.8 ± 0.7	6.0 ± 0.5	1.6 ± 0.5
	F	115.4 ± 0.1	9.0 ± 0.3	1.4 ± 0.1
With VS-treated BF mat	T	58.0 ± 0.7	6.5 ± 0.1	1.5 ± 0.2
	F	109.5 ± 6.4	9.5 ± 0.5	1.3 ± 0.1

The data in Table 7.4 unambiguously demonstrate the BF mat's reinforcing effect as a function of sizing. For mats comprising both as-received and burned BF, the composites' tensile and flexural mechanical properties are equivalent.

Strength and stiffness were significantly increased when VS and ES sized the BF. Keep in mind that the strength properties are significantly impacted by size. However, because stiffness depends on the amount of reinforcement, which is not the case in this instance, it should be indifferent to BF sizing. This can be explained by the fact that the initial IPN structure has been altered, at least close to the fiber surfaces, as a result of the ES and VS size. Accordingly, depending on whether ES or VS scaling was used, BF is probably covered by an EP- or VE-rich layer.

Variations in the crosslinking processes (i.e., VE crosslinking with styrene through a free radical mechanism and polyaddition between the epoxy groups) may have altered the matrix's stiffness and strength properties.

7.8 HYBRID FIBER MAT-REINFORCED HYBRID THERMOSETS

The reinforcing mat, which was 30% present here as well, comprised 70% CeF (10–50 mm long, diameter 9 + 1 mm) and 30% GF (10–50 mm long, diameter 11 + 2µ mm). It just changed this way. The differences between GF and CeF are minimal from a mechanical standpoint. This means that the impact of surface modification may be greater than the impact of fiber hybridization. The results in Table 7.5 suggest that this was most likely the case because the fiber surface treatment had such a significant impact. Once again, as predicted, the VS and ES treatments of the fibers were the most successful in enhancing the performance of the VE/EP hybrid resin-based mat-reinforced composites.

TABLE 7.5
Hybrid Fiber Mat-Reinforced Composites' Tensile and Flexural Mechanical Characteristics

Component	Loading	σ (MPa)	E (GPa)	ε (%)
VE/EP matrix	T	40.8 ± 1.7	2.4 ± 0.1	4.5 ± 0.8
	F	75.9 ± 1.5	2.2 ± 0.1	5.2 ± 0.2
With as-received	T	56.8 ± 1.6	2.7 ± 0.1	4.5 ± 0.3
CeF/GF mat	F	99.3 ± 0.1	2.6 ± 0.1	4.9 ± 0.1
With burned CeF/	T	58.4 ± 1.6	4.8 ± 0.1	2.5 ± 0.2
GF mat	F	119.4 ± 0.1	4.8 ± 0.1	2.9 ± 0.6
With ES-treated	T	86.1 ± 0.2	6.6 ± 0.1	3.4 ± 0.5
CeF/GF mat	F	145.3 ± 4.1	6.4 ± 0.1	2.5 ± 0.1
With VS-treated	T	96.6 ± 0.3	7.6 ± 0.5	3.8 ± 0.2
CeF/GF mat	F	133.1 ± 2.5	6.5 ± 0.1	4.3 ± 0.2

7.9 CONCLUSION

Fiber-reinforced polymer composites are essential in many applications nowadays because of their attractive price/weight characteristics, strong mechanical capabilities, and resistance to corrosion. The findings in this chapter show that melt-blowing-produced discontinuous BF is an effective reinforcement for both thermoplastic and thermoset polymers. However, because of BF's fragility, "soft" preforming and processing processes must be chosen (e.g., carding mats rather than needling). In order to process melts, the machinery may need to be adjusted (for instance, the screw configuration), and the appropriate processing conditions should be chosen. For BF applications, thermoset processing techniques such as reinforced reactive injection molding with polyurethanes, reinforced reactive injection molding with resin transfer molding, and manufacture of sheet molding compounds may be extremely intriguing.

A simple way to lower the cost of materials without losing the mechanical property profile of the composites is by fiber hybridization with BF. In rare circumstances, even a beneficial hybrid effect is possible.

Both thermosets and thermoplastics appear to benefit from matrix hybridization. One of the polymers' encapsulation of BF may have a number of advantageous benefits, including resistance to breaking and customization of the interphase characteristics.

It has been determined that BF surface treatment is essential for achieving useful composite characteristics. But while the BF is still being produced, this issue should be taken into account.

It is important to note that in the not-too-distant future, high-quality continuous and discontinuous BF produced by conventional melt spinning technology will be offered. This might give BF uses in polymer composites a new boost.

Future studies on BF/polymer composites will concentrate on BF's unique qualities, specifically its resilience to fire and corrosion. It is possible to anticipate the use of recyclable polymers with regard to matrix development.

REFERENCES

[1] D. D. Dzhigiris, R. P. Polevoi, and P. P. Polevoi, "Life of refractories of tank furnace for production of superfine basalt fiber," *Glass Ceram.*, 33, 223 (1976).

[2] D. D. Dzhigiris, M. E Makhova, V. D. Gorobinskaya, and V. D. Bombyrlin, "Continuous basalt fiber," *Glass Ceram.*, 40, 467 (1983).

[3] R. V. Subramanian, and H. E Austin, "Silane coupling agents in basalt-reinforced polyester composites," *Int. J. Adhes. Adhes.*, 1, 50 (1980).

[4] J. M. Park, and R. V. Subramanian, "Interfacial shear strength and durability improvement by monomeric and polymeric silanes in basalt fiber epoxy singlefilament composite specimens," *J. Adhes. Sci. Technol.*, 5, 459 (1991).

[5] J. M. Park, R. V. Subramanian, and A. E. Bayoumi, "Interfacial shear strength and durability improvement by silanes in single-filament composite specimens of basalt fiber in brittle phenolic and isocyanate resins," *J. Adhes. Sci. Technol.*, 8, 133 (1994).

[6] I. Wojnirovits, "Factors influencing the mechanical properties of silicate fibers," *Glass Sci. Technol.*, 68, 360 (1995).

[7] T. Jung, and R. V. Subramanian, "Strengthening of basalt fiber by alumina addition," *Scripta Muter.*, 28, 527 (1993).

[8] F. M. Kogan, and O. V. Nikitina, "Solubility of chrysotile asbestos and basalt fibers in relation to their fibrogenic and carcinogenic action," *Environ. Health Persp.*, 102, 205 (1994).

[9] E. E. McConnell, O. Kamstrup, R. Musselman, T. W. Hesterberg, J. Chewalier, W. C. Miller, and P. Thevenaz, "Chronic inhalation study of size-separated rock and slag wool insulation fibers in Fischer 344/N rats," *Inhal. Toxicol.*, 6, 571 (1994).

[10] Z. Adamis, T. Kerknyi, K. Honma, M. Jackel, E. Titrai, and G. Ungviry, "Study of inflammatory responses to crocidolite and basalt wool in the rat lung," *J. Toxicol. Env. Health*, 62, 409 (2001).

[11] J. M. Park, E. M. Chong, W. G. Shin, S. I. Lee, D. J. Yoon, and J. H. Lee, "Interfacial properties and micro-failure mechanisms of SiC fiber reinforced high temperature thermoplastic composites using fragmentation technique and acoustic emission," *Polym. Korea*, 20, 753 (1996).

[12] J. M. Park, E. M. Chong, D. J. Yoon, and J. H. Lee, "Interfacial properties of two SiC fiber-reinforced polycarbonate composites using the fragmentation test and acoustic emission," *Polym. Compos.*, 19, 747 (1998).

[13] J. M. Park, and W. G. Shin, "Interfacial aspects of dual basalt and SiC fibers reinforced epoxy composites using fragmentation technique and acoustic emission," *Korea Polym. J.*, 5, 114 (1997).

[14] J. M. Park, W. G. Shin, and D. J. Yoon, "A study of interfacial aspects of epoxybased composites reinforced with dual basalt and SiC fibres by means of the fragmentation and acoustic emission techniques," *Compos. Sci. Technol.*, 59, 355 (1999).

[15] P. I. Bashtannik, V. G. Ovcharenko, and Y. A. Boot, "Effect of combined extrusion parameters on mechanical properties of basalt fiber-reinforced plastics based on polypropylene," *Mech. Compos. Matel.*, 33, 600 (1997).

[16] P. I. Bashtannik, and V. G. Ovcharenko, "Antifriction basalt-plastics based on polypropylene," *Mech. Compos. Matel.*, 33, 299 (1997).

[17] M. Botev, H. Betchev, D. Bikiaris, and C. Panayiotou, "Mechanical properties and viscoelastic behavior of basalt fiber-reinforced polypropylene," *J. Appl. Polym. Sci.*, 74, 523 (1999).

[18] W. B. Goldsworthy, "New basalt fiber increases composite potential," *Compos. Technol.*, 8, 15 (2000).

[19] V. V. Gurev, E. I. Neproshin, and G. E. Mostovoi, "The effect of basalt fiber production technology on mechanical properties of fiber," *Glass Ceram.*, 58, 62 (2001).

[20] N. N. Morozov, V. S. Bakunov, E. N. Morozov, L. G. Aslanova, P. A. Granovskii, V. V. Prokshin, and A. A. Zemlyanitsyn, "Materials based on basalts from the European North of Russia," *Glass Ceram.*, 58, 100 (2001).

[21] A. M. Zihlif, and G. Ragosta, "A study on physical properties of rock wool fiber-polystyrene composite," *J. Themoplast. Compos.*, 16, 273 (2003).

[22] P. I. Bashtannik, A. I. Kabak, and Y. Y. Yakovchuk, "The effect of adhesion interaction on the mechanical properties of thermoplastic basalt plastics," *Mech. Compos. Matel.*, 39, 85 (2003).

[23] K. Friedrich, and S. Fakirov, and Z. Zhang, *Polymer Composites: From Nano- to Macro-Scale.* Springer Science & Business Media, 2005.

[24] J. Karger-Kocsis, "Reinforced polymer blends," in D. R. Paul, and C. B. Bucknall, eds., *Polymer Blends*, John Wiley & Sons, New York, Vol. 2, p. 395, 2000.

[25] S. Y. Fu, G. Xu, and Y. W. Mai, "On the elastic modulus of hybrid particle shortfiber polymer composites," *Compos. Part A*, 33, 291 (2002).

[26] B. Pukinszky, "Particulate filled polypropylene: Structure and properties," in J. Karger-Kocsis, ed., *Polypropylene: Structure, Blends and Composites*, Chapman & Hall, London, Vol. 3, p. 1, 1995.

[27] A. R. Bunsell, and B. Harris, "Hybrid carbon and glass fiber composites," *Composites*, 4, 157 (1974).

[28] J. Summerscales, and D. Short, "Carbon fibre and glass fibre hybrid reinforced plastics," *Composites*, 9, 157 (1978).

[29] G. Marom, E. Drukker, A. Weinberg, and J. Banbaji, "Impact behaviour of carbon/Kevlar hybrid composites," *Composites*, 17, 150 (1986).

[30] G. Marom, H. Harel, S. Neumann, K. Friedrich, K. Schulte, and H. D. Wagner, "Fatigue behaviour and rate-dependent properties of aramid fibrelcarbon fibre hybrid composites," *Composites*, 20, 537 (1989).

[31] R. A. Chaudhuri, and H. J. Garala, "Analytical experimental evaluation of hybrid commingled carbon glass epoxy thick-section composites under compression," *J. Compos. Matel.*, 29, 1695 (1995).

[32] S. Y. Fu, B. Lauke, E. MBder, X. Hu, and C. Y. Yue, "Fracture resistance of short glass-fiber-reinforced and short-carbon-fiber-reinforced polypropylene under Charpy impact load and its dependence on processing," *J. Matel. Process. Tech.*, 89–90, 501 (1999).

[33] J. Jang, and C. Lee, "Performance improvement of GFICF functionally gradient hybiid composite," *Polym. Test.*, 17, 383 (1998).

[34] M. Sohn, and X. Hu, "Processing of carbon-fibre epoxy composites with costeffective interlaminar reinforcement," *Compos. Sci. Technol.*, 58, 211 (1998).

[35] K. Friedrich, and O. Jacobs, "On wear synergism in hybrid composites," *Compos. Sci. Technol.*, 43, 71 (1992).

[36] G. Kalaprasad, S. Thomas, C. Pavithran, N. R. Neelakantan, and S. Balakrishnan, "Hybrid effect in the mechanical properties of short sisal/glass hybrid fiber reinforced low density polyethylene composites," *J. Reint Plast. Comp.*, 15, 48 (1996).

[37] G. Kalaprasad, K. Joseph, and S. Thomas, "Influence of short glass fiber addition on the mechanical properties of sisal reinforced low density polyethylene composites," *J. Compos. Mater.*, 31, 509 (1997).

[38] O. I. Benevolenski, J. Karger-Kocsis, K. P. Mieck, and T. Reussmann, "Instrumented perforation impact response of polypropylene composites with hybrid reinforcement flax/glass and flax/cellulose fibers," *J. Thermoplast. Compos. Mater.*, 13, 481 (2000).

[39] H. D. Rozman, G. S. Tay, R. N. Kumar, A. Abusamah, H. Ismail, and Z. A. Mohd Ishak, "Polypropylene-oil palm empty fruit bunch-glass fibre hybrid composites: A preliminary study on the flexural and tensile properties," *Eur. Polym. J.*, 37, 1283 (2001).

[40] J. Rout, M. Misra, S. S. Tripathy, S. K. Nayak, and A. K. Mohanty, "The influence of fibre treatment on the performance of coir-polyester composites," *Compos. Sci. Technol.*, 61, 1303 (2001).

[41] M. M. Thwe, and K. Liao, "Effects of environmental aging on the mechanical properties of bamboo-glass fiber reinforced polymer matrix hybrid composites," *Compos. A*, 33, 43 (2002).

[42] S. Mishra, A. K. Mohanty, L. T. Drzal, M. Misra, S. Parija, S. K. Nayak, and S. S. Tripathy, "Studies on mechanical performance of biofibrelglass reinforced polyester hybrid composites," *Compos. Sci. Technol.*, 63, 1377 (2003).

[43] G. Marom, S. Fisher, F. R. Tuler, and H. D. Wagner, "Hybrid effects in composites: Conditions for positive or negative effects versus rule of mixture behaviour," *J. Mater. Sci.*, 13, 1419 (1978).

[44] S. Fu, Y. Mai, B. Lauke, and C. Yue, "Synergistic effect on the fracture toughness of hybrid short glass fiber and short carbon fiber reinforced polypropylene composites," *Mat. Sci. Eng.*, 323, 326 (2002).

[45] T. Hayashi, "Development of new material properties by hybrid composition," *Compos. Mater.*, 1, 18 (1972).

[46] X. Zhu, Z. Li, and Y. Jin, "Creep behaviour of a hybrid fibre (glass1carbon)-reinforced composite and its application," *Compos. Sci. Technol.*, 50, 431 (1994).

[47] C. H. Chiu, K. H. Tsai, and W. J. Huang, "Crush-failure modes of 2D triaxially braided hybrid composite tube," *Compos. Sci. Technol.*, 59, 1713 (1999).

[48] Y. Shan, and K. Liao, "Environmental fatigue of unidirectional glass-carbon fiber reinforced hybrid composite," *Compos. B*, 32, 355 (2001).

[49] J. R. Hazen, "Carbon fiberglass wind blades planned," *High Performance Compos.*, 11, 13 (2003).

8 Natural Fiber Composites

8.1 INTRODUCTION

Consumer awareness of new items made from renewable sources has increased significantly during the past few years. Consumers have been directed toward ecologically friendly outcomes by green marketing, new perspectives on recycling, social influence, and changes in cognitive values. To offer new products responsibly and sustainably while enhancing and adapting existing ones, composite materials are specifically being developed and modified.

Natural fibers are made by processes in plants, animals, and the earth's crust. They can be utilized as a part of composite materials, where the qualities are affected by the orientation of the fibers. A few effective natural fiber evolutions are detailed in the Table 8.1 below and contrasted with E-glass fiber in the property.

One of the first crucial choices to be made when constructing composite parts is the choice of reinforcement. Natural fibers add sustainability to the whole production, so you shouldn't discount them as a possibility when selecting the best reinforcement for your application until you have exhausted all other options. However, there are a few crucial characteristics that you should think about before choosing. A few benefits and relative drawbacks of natural fibers are noted in the Table 8.1 below. The properties of natural fibers are provided in study [1].

Natural fiber polymer composites (NFPC) are composite materials made of high-strength natural fibers incorporated in a polymer matrix. The pros and cons of natural fibers are provided in study [1].

The polymers that bind the natural fibers together are another important part of NFPCs. Thermoplastics and thermosets are typically the two basic groups into which polymers can be divided for structural applications.

Because the structure of thermoplastic matrix materials is made up of one- or two-dimensional molecules, these polymers have a tendency to become softer at higher temperatures and to regain their original qualities as they cool. Contrarily, thermosets are strongly cross-linked polymers that can either be cured by heat alone or by heat, pressure, and light irradiation. Due to their structure, thermoset polymers have excellent strength and modulus as well as considerable flexibility for customizing desired final qualities. Thermoplastics are frequently used for NFPCs because they may be recycled numerous times, as opposed to thermoset polymers, which must either undergo a chemical procedure or be burned to dispose them.

8.2 CHARACTERISTICS

Each natural fiber composite is unique in its characteristics, because of various sources, such as moisture conditions, and fiber types, as mentioned in earlier studies. The mechanical composition, microfibrillar angle, structure, flaws, cell

 DOI: 10.1201/9781003429197-8

diameters, physical properties, chemical properties, and the interaction of a fiber with the matrix are all important determinants of NFPC performance.

Natural fiber-reinforced polymer composites have limitations, just like every other product on the market does. Due to the differences in chemical structure between these two phases, the joints between natural fiber and polymer matrix constitute a concern. It results in inefficient stress transfer at the NFPCs' interface.

Natural fibers are hydrophilic because they contain an active component called the hydroxyl group. Due to the hydroxyl group in natural fibers, there is poorer interfacial interaction between hydrophilic natural fiber and hydrophobic polymer matrices in NFPCs. It might result in NFPCs with supple mechanical and physical characteristics.

8.3 FACTORS AFFECTING NFPC'S

The characteristics and performance of NFPCs can be influenced by a variety of factors. The hydrophilicity of natural fibers and their loading have an effect on the composite materials' characteristics as well. To achieve the desirable features of NFPCs, substantial fiber loading is typically required.

Observe that the tensile characteristics of composites generally improve as fiber content increases. The process parameters used are another important component that has a significant impact on the properties and surface features of the composites. To produce composites with the greatest qualities, one must carefully select the right process parameters and processes. The features of the composite, which are represented by the percentage of cellulose, hemicellulose, lignin, and waxes, are also greatly influenced by the chemical composition of natural fibers.

8.4 APPLICATIONS OF NFPC

Natural fibers are being used more frequently in a variety of industries, including construction, furniture, packaging, and transportation. This is primarily because they have advantages over synthetic fibers, such as low cost, low weight, less harm to processing equipment, improved surface quality of composite parts, good relative mechanical qualities, and abundant and renewable resources. Natural fibers are utilized in a variety of products, including clothing, biopolymers, fine chemicals, food, feed for humans and animals, particle boards, insulation, construction materials, and building materials.

8.4.1 AUTOMOTIVE BUSINESSES

Due to the fact that the side and front panels of vehicles are not primary structural components, NFPC composites are acceptable in the automobile industry. When NFPC composites are used in these components in place of conventional glass fiber composites and aluminum, the price and weight of automobiles may be decreased in part.

8.4.2 CONSTRUCTION-RELATED INDUSTRIES

Natural fiber-cement composites have only been used for exterior components of residential buildings, such as siding and roofing. In order to preserve these structures from weathering assault, additional protective coating material is required.

8.4.3 FORCING ADOPTION THROUGH FUNCTION

Governments and society are putting pressure on a number of businesses, including the automotive, energy, construction, and aerospace sectors, to produce more ecologically friendly goods and lessen their reliance on fossil fuels. The European Commission implemented "European Guideline 2000/53/EG" to increase automotive reusability to 85% for a vehicle by weight in 2005. By 2015, this ratio had increased to 95%.

Adoption of natural fiber composites is significantly influenced by this kind of codification. In this situation, natural fibers are a desirable option for enterprises to address socioeconomic and environmental concerns.

Additionally, the use of natural fibers would generate employment opportunities in rural and underdeveloped areas, aiding in the achievement of the United Nations' realistic development goals, including reducing poverty, fostering innovation, achieving comprehensive and sustainable industrialization, building sustainable cities and communities, and promoting responsible production and consumption. Therefore, natural fibers will be essential to the socioeconomic growth of our society.

8.5 NATURAL FIBER COMPOSITE (NFC) DESIGN

Industrial design (ID) is the disciplined activity of creating things that are utilized every day by millions of people all over the world. One of the divisions of industrial design that includes home appliances, safety gear, and medical equipment is product design. Before moving on to the production stage, a successful product design must go through a protracted procedure called the design process. Study [2] claims that the field of furniture design is extremely broad and encompasses furniture for private, public, and commercial areas. Transportation design includes all land-, water-, and air-based vehicles, including cars, motorcycles, buses, sea trucks, ships, jet skis, helicopters, and airplanes. According to Abidin et al. [3], the industrial sector in the United States produced more than 1235×10^6 metric tons of carbon dioxide gas in 2007. As a result, the restoration of greenhouse gases would be more challenging.

8.6 MATERIAL SELECTION

Product designers may use natural fiber composite materials in their design concepts to promote their creations as eco-designs. Eco-design, often known as design for the environment, is the process of "integrating a systematic environmental

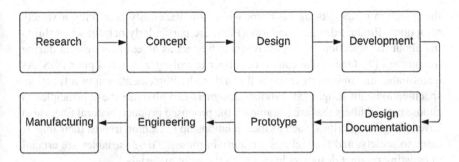

FIGURE 8.1 Designing process.

system into product design and development" [4]. Designers need to be very careful when deciding whether natural fiber is ideal for a given product. According to a study by Yung et al. [5], the choice of material frequently depends on the material that has already been used in order to ensure that the material being used is secure. However, this approach limits the range of materials available. An innovative product is produced in large part thanks to the material choice.

According to a study by Yung et al. [5], materials can be distinguished based on their mechanical properties like tensile strength and thermal conductivity as well as their sensory characteristics, such as blurred texture and transparency, or their ability to be processed and shaped, like being painted or injected. The conceptual design phase is where this procedure must be finished. Figure 8.1 depicts the design process from the original idea drawing stage to the 2D rendering stage [6].

Designers should therefore develop an overview of these essential characteristics, including sensorial characteristics (such as transparency and velvet-like texture), technical characteristics (such as its specific tensile strength), and formability characteristics, in order to motivate and inspire consumers to choose a particular material (such as its ability to undergo injection molding or being paintable). Unless specific technical requirements are established at the project's outset, product designers only consider technical attributes at an overview level during the conceptual design stage [5]. According to Marzuki [7], designers can use the Meanings of Material (MoM) model as a guide when selecting materials. Product designers can also use these principles to choose natural fibers that are suitable for the proposed product. Natural fibers also have intangible qualities, such as their relationship to fashion trends, their importance to society, and the feelings a material arouses. These variables are crucial in assisting product designers in choosing the right materials.

8.7 CONCEPT DESIGN

For the evaluation and selection of their design concepts, product designers require trustworthy, exacting, and strong methodologies. Selecting the proper approach is crucial, and selecting the incorrect product design proposal can cost the company a lot of money, effort, and other vital resources. Designers must continually assess the direction of their design concept while simultaneously producing a variety of options. Product design standards (PDS) are particularly helpful when thinking about the selection of proposals since they act as assessment criteria during the process [2]. One of the causes of subpar or failing goods is a poor PDS. As quantifiable dimensions of features that aid in the implementation of a function, specifications are sought [8]. Product designers can also use these principles to choose natural fibers that are suitable for the proposed product. Natural fibers also have intangible qualities, such as their relationship to fashion trends, their importance to society, and the feelings a material arouses. These variables are crucial in assisting product designers in choosing the right materials.

The Paugh method, also known as matrix assessment, is a quantitative tool that designers use to score their suggested design concepts against predetermined standards laid down in the PDS. This technique was developed by Stuart Paugh, a British engineering design professor who is regarded as a pioneer in the field of product design development, and it is now used extensively in the field of design for production. One of the procedures for reducing the number of potential solutions is the selection of concept design ideas, which attempts to choose one for additional development and improvement. The choice matrix model (Paugh's technique) is shown below. Mahmud et al. [9] contend that a significant increase in the amount of information available on product design specifications (PDS) throughout the design phase has led to a significant drop in the occurrence of product desertion. It is important to adhere to the PDS for selecting the most satisfying design proposal if you want to ensure that the suggested design idea does not deviate from the restrictions laid out in the PDS. Matrix evaluation may be helpful to designers, engineers, manufacturers, marketing professionals, users, customers, and purchasers by lowering ambiguity and confusion in the assessment and selection process. By doing this, successful new items will be introduced to the market more frequently and with clearer communication.

8.8 INCORPORATING SUSTAINABLE DESIGN WITH OTHER CONCURRENT ENGINEERING PROCESSES DURING PRODUCT DEVELOPMENT

To promote sustainable products, product engineers must apply the concept of design for sustainability (DfS) to the development of NFC items. DfS, which is defined in terms of the four sustainability pillars (Figure 8.2) that are required to achieve sustainable life quality: (1) ecological, (2) social, (3) economic, and (4) institutional, could be a key factor in guiding us toward sustainable consumption and production. Designers should think about and assess these four pillars from obtaining resources to producing completed things in order to practice sustainability [10]. Including elements that comply with consumption and production standards, including the use of the best technology, materials, and manufacturing procedures, is crucial if one is to attain zero carbon emissions and the least amount of non-renewable resource use [10]. The effects on human well-being must also be taken into account.

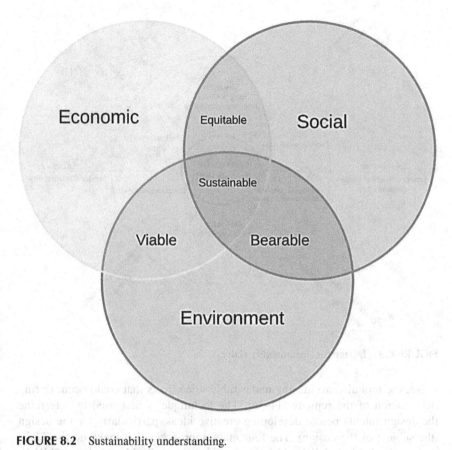

FIGURE 8.2 Sustainability understanding.

Typically, the DfS method uses design for excellence (DfX) to create a sustainable product. In order to produce sustainable components and products, this approach entails analyzing the environmental effects of certain design qualities, including safety and biodegradability possibilities. Prior to the manufacturing process, concurrent engineering must be included in product development in order to meet customer expectations and create sustainable goods, according to Spangenberg et al. [11]'s conceptual framework for DfS (Figure 8.3). A biocomposite product must adhere to life cycle analysis and sales trends, taking into account factors like the cost of raw materials and manufacture, the performance of the product, and customer needs [12,13].

8.9 THEORY OF INVENTIVE PROBLEM-SOLVING (TRIZ)

Concurrent engineers employ TRIZ, or the idea of imaginative problem solving, as a tool to create a variety of solutions based on inventive principles to address problems as they emerge [13]. Due to its emphasis on the problem's fundamental

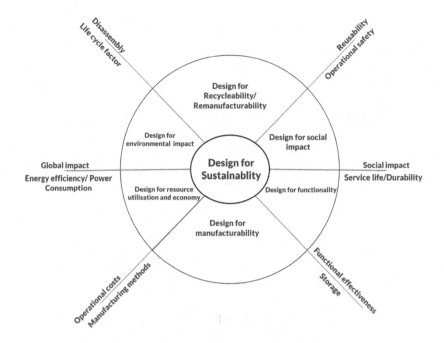

FIGURE 8.3 Design for sustainability (Dfs).

cause, the tool also avoids any undesirable side effects that could occur during the creation of the remedy [14–16]. The technique is first used to determine the design intents before developing creative ideas, particularly for the design (the subject of the design). The four main approaches that make up the TRIZ tool are Su-field modeling, Algorithms of Ingenious Problem Solving (ARIZ), Technology Trend Prediction, and Contradiction Engineering with 40 Ingenious Principles [17]. The application of opportunity and innovation methods depends on how complicated the issues being addressed are in order to effectively address them [18]. In this case, Rosli et al. [19] used TRIZ contradiction ways to develop a fresh concept for sheet metal snips. These procedures compare parameters that are improving and deteriorating to select relevant innovative concepts. Toward the end of the product development period, they enhanced the design concepts using a CAD optimization tool. Figure 8.4 [20] depicts the design process used by Asyraf.

8.9.1 Customer Voice

One of the key strategies used in concurrent engineering to create ideas for design intentions is the voice of the customer (VOC). Direct consumer specifications, observation, surveys, group discussions, interviews, focus groups, warranty information, and field reports are just a few of the methods used to collect customer feedback. These VOC data are then used to inform the deployment of a quality

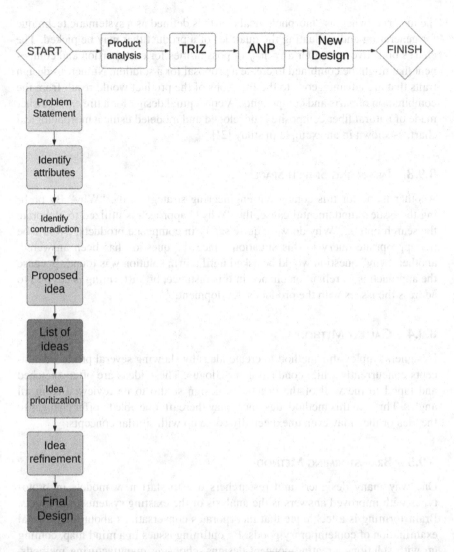

FIGURE 8.4 Theory of inventive problem solving (TRIZ).

function or a product planning matrix (QFD). The QFD is used to categorize client demands and put these data into methodical strategies for producing goods that satisfy those needs.

8.9.2 Chart for Morphology

The morphological chart is a concurrent engineering tool that employs a chart with several arrangements to assist designers in selecting new groupings of characteristics or parts. "Morphology" refers to the study of the form or shape of the material, whereas "morphological chart" is defined as a systematic technique for generating and evaluating the qualities of a product that may be picked. The chart's objective is to offer a variety of possibilities for each portion and component that might be combined to create a proposal for a solution. Numerous design traits that are advantageous to the functions of the product would result from the combination of parts and components. A conceptual design for a fire extinguisher made of natural fiber composite is developed and modeled using a morphological chart, as shown in an example in study [21].

8.9.3 INCREASING SEARCH SPACE

Another name for this concurrent engineering strategy is the "Why? By probing the issue's fundamental cause, the "Why?" approach is utilized to elaborate the search option. "Why do we require safety in composite products?" would be the appropriate query in this situation. Once the question has been answered, another "why" question would be asked until a firm solution was found. Because the approach is so reliant on chance in this instance, brainstorming is advised to address the issues with the product's development.

8.9.4 GALLERY METHOD

Designers employ this method to create ideas by showing several produced concepts concurrently while conducting a dialogue. These ideas are often sketched and taped to the wall of the designer's design studio to be reviewed from all angles. Through this method, designers may thereafter be able to offer changes to the idea or they may even unexpectedly come up with similar concepts.

8.9.5 BRAINSTORMING METHOD

One way many designers and researchers use to start new models or prototypes with improved answers is the analysis of the existing systems or products. Brainstorming is a technique that incorporates conversations about the physical examination of contemporary goods. By outlining issues in a mind map, coming up with solutions, creating concept designs, choosing manufacturing methods, and creating finished prototypes, the debate may result in a clearer image. The examination of an existing product is typically done in comparison to similar goods that have several sub-functions of function structures, earlier products of the same firm, and products from rivals.

8.10 APPLICATIONS

In several technical domains, NFC applications are expanding quickly. Corn, water hyacinth, coir, ginger, cotton, kenaf, sugarcane, flax, ramie, hemp, kapok, sisal, wood, oil palm, banana, as well as sugar palm have all been utilized as reinforcements in polymer composites. In addition to being biodegradable, natural fibers have a number of other benefits, such as being less expensive, readily available, and preventing deforestation. Ahmad et al.'s [22] discovery that natural fibers are the ideal material for the replacement of glass and carbon fibers suggests that natural fibers have enormous potential to be turned into valuable goods.

Different automotive applications, structural components, packaging applications, furnishings, and structures have come to place a high value on various natural fibers including jute, hemp, kenaf, oil palm, and bamboo-reinforced polymer composites. NFCs are utilized in the electrical and electronic sectors as well as in aerospace, sporting goods, leisure equipment, watercraft, manufacturing, office supplies, and other industries. Study [23] illustrates how San et al. [23] created a design for sustainability (DfS)-based roselle fiber-reinforced polymer composite smartphone holder. Idea creation and concept assessment approaches were used to generate the concept for the eco-friendly smartphone holder device. Market research, the preparation of product design specifications (PDS) documents, the creation of conceptual designs, and the detailed design of the final product were all steps in the development process for the Roselle Composite Smartphone Holder. A 3D printing process was used to create the product's mold. Then, using a hand layup technique, the roselle fiber composite smartphone holder was created.

Study [24] illustrates the results of another study carried out by Ilyas et al. on the product development process of a roselle fiber-reinforced polymer composite mug pad utilizing the sustainability (DfS) method. Idea creation and concept assessment approaches were used to generate the concept for the ecologically friendly mug pad product. They used comparable procedures to San et al. work [23], in which the molded product's final design was created using a 3D printer and the roselle fiber composite mug pad was created using a manual layup procedure. The finished design was simple to fabricate, lightweight, affordable overall, and struck the right mix between utility and aesthetics.

8.10.1 NFC Applications in Electronic Components

Due to the increasing importance of raw materials generated from renewable resources and the recyclability or biodegradability of products, natural fibers are currently replacing petroleum-based synthetics in electrical and electronic applications. Natural fiber-reinforced composites are advantageous for a range of uses in the electrical and electronic industries due to their low weight, high stiffness-to-weight ratio, and biodegradability. The "FOMA(R) N701iECOeco-phone" cover is built of kenaf fibers, as seen in study [5].

8.10.2 NFC Applications in Packaging

Natural fiber composites have lately offered an alternative for improved packaging. Prior to recently, the majority of petroleum-based plastics used to package food were non-degradable, which resulted in a number of environmental issues when they were disposed of, including harm to the environment and eco-systems, water sources, sewage systems, rivers, and streams. In addition, they are non-renewable, and because of the upcoming depletion of petroleum supplies, their costs are also growing and unstable. To use less non-renewable and petroleum-based resources, coir (coconut) fiber reinforced with natural latex is being used more and more often in favor of synthetic materials. Coconut fiber is an exceptionally resilient and elastic substance that virtually ever degrades over time. It is a strong substance that may be used repeatedly. It may easily be recycled or disposed of after being used. The substance is heated to vulcanize the natural latex after shaping it into the desired form. The end product is a sturdy and durable construction that is quite open. The packing items made from coconut fiber by Enkev Manufacturer are seen in study [16].

8.10.3 NFC Applications in Sports Equipment

Although the automobile sector is where natural fiber composites are most frequently used, there are other industries where they are also used, such as sports equipment. Prior to the invention of fiber-reinforced composites, materials for sporting goods were made of wood, steel, stainless steel, aluminum, and alloys. The following are some ways in which fiber-reinforced composite materials obviously outperform these materials. Natural fibers' smaller weight and similarly lower cost are frequently stated as the main arguments in favor of their use in composite sporting products. Since most sporting goods are moved by people, lightweight equipment is recommended. According to a study, an oil palm empty fruit bunch fiber/epoxy composite that underwent a 24-hour fiber treatment held a great deal of promise for use as a reinforcement to epoxy as a suitable material for sporting goods. Based on the information gathered related to the mechanical and physical characteristics, the composite of OPEFB fiber/epoxy has a flexural strength between 67.90 and 83.63 MPa, which is within the range of the strength requirements for field hockey sticks.

8.11 CONCLUSION

Despite the fact that a variety of parameters influence the usage of natural fibers as reinforcement in polymer-based composites, their cost-competitiveness and renewability continue to entice businesses from all industries to look for ways to replace conventional materials with natural fibers. Given the vast number of problems that remain unsolved, study in this area is extremely valuable. It goes without saying that properly designing NFRP composites and selecting an appropriate production method will help them become one of the key structural materials in the engineering sectors in the future.

This chapter presents a clear and instructive review of natural fiber-reinforced polymer matrices from the perspective of product design development. Plant fibers, often referred to as cellulosic fibers, are one of the three main sources of natural fibers, along with animals and minerals, and are in high demand since they are non-toxic to humans, animals, and the environment. As their fundamental building ingredients, cellulose, hemicellulose, lignin, and pectin are commonly found in natural fibers. Numerous investigations have demonstrated that these natural fibers perform well mechanically because cellulose gives them their excellent shape and structural stability. As a result, natural fiber and polymer composites have a number of advantages over synthetic composites, including decreased solidity, reduced density, biodegradability, and lower cost. Natural fiber composites are a practical way to improve the quality of products manufactured from them in terms of environmental acceptability as well as technical and economic feasibility. The most often used natural fibers in composite goods are flax, coir, hemp, and jute. Due to their high mechanical strength and stiffness, which are suitable for a variety of technical applications, kenaf, sugar palm, and roselle are examples of developing fibers. Effective product design and production techniques of NFPCs are required to increase the attributes of the goods and their materials toward maximal strength and utility. The voice of the customer (VOCs), brainstorming, TRIZ, and morphological charts are some of the engineering design processes and approaches that are essential to ensure the strength and utility of natural fiber composite products are maximized. These techniques could identify user problems and explain them in terms of how the product works. In the end, a suitable production method incorporates the product's design and intended usage. To develop engineering design tactics that are ideal for production procedures, heavy industry applications, and the strength of natural fiber composites, further study will be required. Numerous applications that do not require exceptionally strong load-bearing or high-temperature operating characteristics may currently use natural fiber composites.

REFERENCES

[1] D. B. Dittenber, and H. V. S. GangaRao, "Critical review of recent publications on use of natural composites in infrastructure," *Compos. Part A*, 43, 1419–1429 (2012).

[2] What Is Industrial Design? [(accessed on 18 March 2021)]; Available online: https://www.idsa.org/what-industrial-design.

[3] S. Z. Abidin, M. H. Abdullah, and Z. Yusoff, *Seni reka perindustrian: Daripada Idea Kepada Lakaran Dewan Bahasa dan Pustaka*, Seni Reka Perindustrian, Kuala Lumpur, Malaysia, 2013.

[4] K. Ramani, D. Ramanujan, W. Z. Bernstein, F. Zhao, J. Sutherland, C. Handwerker, J.-K. Choi, H. Kim, and D. Thurston, "Integrated sustainable life cycle design: A review," *J. Mech. Des.*, 132, 1–15 (2010). doi:10.1115/1.4002308.

[5] W. K. Yung, H. K. Chan, J. H. So, D. W. Wong, A. C. Choi, and T. M. Yue, "A life-cycle assessment for eco-redesign of a consumer electronic product," *J. Eng. Des.*, 22, 69–85 (2011). doi:10.1080/09544820902916597.

[6] E. Karana, "Materials selection in design: From research to education," *Proceedings of the the 1st International Symposium for Design Education Researchers*, Paris, France, pp. 18–19, 2011.

[7] I. Marzuki, *Reka Bentuk Produk,* Dewan Bahasa dan Pustaka, Kuala Lumpur, Malaysia, Proses Reka Bentuk Produk, p. 16, 2013.

[8] E. Karana, P. Hekkert, and P. Kandachar, "A tool for meaning driven materials selection," *Mater. Des.*, 31, 2932–2941 (2010). doi:10.1016/j.matdes.2009.12.021.

[9] J. Mahmud, S. Khor, M. M. Ismail, J. M. Taib, N. Ramlan, and K. Ling, "Design for paraplegia: Preparing product design specifications for a wheelchair," *Technol. Disabil.*, 27, 79–89 (2015). doi:10.3233/TAD-150430.

[10] M. A. Azman, S. A. M. Yusof, I. Abdullah, I. Mohamad, and J. S. Mohammed, "Factors influencing face mask selection and design specifications: Results from pilot study amongst Malaysian Umrah pilgrims," *J. Teknol.*, 79, 79 (2017). doi:10.11113/jt.v79.9779.

[11] J. H. Spangenberg, A. Fuad-Luke, and K. Blincoe, "Design for sustainability (DfS): The interface of sustainable production and consumption," *J. Clean. Prod.*, 18, 1485–1493 (2010). doi:10.1016/j.jclepro.2010.06.002.

[12] J. H. Spangenberg, "Sustainable development indicators: Towards integrated systems as a tool for managing and monitoring a complex transition," *Int. J. Glob. Environ. Issues*, 9, 318 (2009). doi:10.1504/IJGENVI.2009.027261.

[13] M. A. G. Von Keyserlingk, N. P. Martin, E. Kebreab, K. F. Knowlton, R. J. Grant, M. Stephenson, C. J. Sniffen, J. R. Harner, III, A. D. Wright, and S. I. Smith, "Invited review: Sustainability of the US dairy industry," *J. Dairy Sci.*, 96, 5405–5425 (2013). doi:10.3168/jds.2012-6354.

[14] I. S. Jawahir, K. E. Rouch, O. W. Dillon, L. Holloway, A. Hall, and J. Knuf, "Design for sustainability (DFS): New challenges in developing and implementing a curriculum for next generation design and manufacturing engineers," *Int. J. Eng. Educ.*, 23, 1053–1064 (2007).

[15] A. Hambali, S. M. Sapuan, N. Ismail, and Y. Nukman, "Application of analytical hierarchy process in the design concept selection of automotive composite bumper beam during the conceptual design stage," *Sci. Res. Essays*, 4, 198–211 (2009).

[16] N. Mazani, S. Sapuan, M. Sanyang, A. Atiqah, and R. Ilyas, "Design and fabrication of a shoe shelf from kenaf fiber reinforced unsaturated polyester composites," in *Lignocellulose for Future Bioeconomy*, Elsevier, Amsterdam, The Netherlands, pp. 315–332, 2019.

[17] G. Pahl, and W. Beitz, *Engineering Design: A Systematic Approach*, Springer, London, UK, 1996.

[18] M. Asyraf, M. Ishak, S. Sapuan, and N. Yidris, "Conceptual design of multi-operation outdoor flexural creep test rig using hybrid concurrent engineering approach," *J. Mater. Res. Technol.*, 9, 2357–2368 (2020). doi:10.1016/j.jmrt.2019.12.067.

[19] M. U. Rosli, M. K. A. Ariffin, S. M. Sapuan, and S. Sulaiman, "Integrated AHP-TRIZ innovation method for automotive door panel design," *Int. J. Eng. Technol.*, 5, 3158–3167 (2013).

[20] M. Li, X. Ming, L. He, M. Zheng, and Z. Xu, "A TRIZ-based trimming method for patent design around," *Comput. Des.*, 62, 20–30 (2015). doi:10.1016/j.cad.2014.10.005.

[21] M. R. M. Asyraf, M. Rafidah, M. R. Ishak, S. M. Sapuan, N. Yidris, R. A. Ilyas, and M. R. Razman, "Integration of TRIZ, morphological chart and ANP method for development of FRP composite portable fire extinguisher," *Polym. Compos.*, 41, 2917–2932 (2020). doi:10.1002/pc.25587.

[22] S. A. Ahmad, M. C. Ang, K. W. Ng, and A. N. Abdul Wahab, "Reducing home energy usage based on TRIZ concept," *Adv. Environ. Biol.*, 9, 6–11 (2015).

[23] Y. T. San, Y. T. Jin, and S. C. Li, *TRIZ: Systematic Innovation in Manufacturing*, Firstfruit Sdn. Bhd., Selangor, Malaysia, 2011.

[24] T. Li, "Retracted article: Applying TRIZ and AHP to develop innovative design for automated assembly systems," *Int. J. Adv. Manuf. Technol.*, 46, 301–313 (2009). doi:10.1007/s00170-009-2061-4.

9 Vibration and Noise

9.1 COMPOSITE MATERIALS IN THE AUTOMOTIVE INDUSTRY

Composite materials are being used more frequently in place of metallic components in the automotive industry. For automotive parts such as body panels, chassis components, and engine parts, the use of sheet molding compound (SMC), which is based on liquid unsaturated polyester (UP), has risen. The composition of SMC frequently includes UP resin (25%–30%), fiber reinforcing material (25%–30%), mineral filler (40%–45%), and extra additives like a release agent, color, thickening agent, and so on. A low-pressure moldable SMC called LPMC was introduced by the Scott Bader Company. Low-pressure molding compound is a new type of composite material that can be used for automotive body panels (LPMC). Because of a special thickening mechanism, LPMC offers great moldability and shares mechanical traits with conventional SMC. The Scott Bader Company uses a novel thickening technique that replaces the metal oxide included in a normal SMC formulation with crystalline polyester, also referred to as Crystic Impreg. The crystalline resin is dissolved by the liquid resin at 82°C, which is then transformed into a paste and made into LPMC while maintaining a temperature above 50°C. In other situations, like loading glass fiber and feeding resin, standard SMC is fairly equivalent. The crystalline resin no longer dissolves and the viscosity nearly increases quickly when a substance is converted into LPMC and then allowed to cool to ambient temperature. In contrast to metal oxide thicker SMC, additional maturation is therefore not necessary, and the handling qualities last far longer. The primary cause of LPMC's advantages is its unique thickening procedure, which gives it flow qualities that are superior to those of conventional SMC. Furthermore, compression molding of LPMC materials is possible at substantially lower pressures (1–3 MPa), which reduces the expense of tooling and maintenance.

9.2 UNDERSTANDING OF MULTILAYER COMPOSITE MATERIALS

In a number of circumstances, noise and vibration can affect how comfortable car passengers are. Automotive interiors with lower vibration and noise levels are regularly used as standards for product acceptance. Engineers therefore put a lot of effort into developing a tranquil environment that would satisfy clients. The vibration management components are utilized in vehicle design,

DOI: 10.1201/9781003429197-9

such as absorbers, barriers, dampers, and isolators. Surface-damping materials, which are useful for reducing structure-borne noise at frequencies higher than 100 Hz, are modeled and examined in this section. These methods of surface dampening significantly reduce vibrations in structural panels. The door, roof, dash, floor, and cab rear panels of the car are typically coated with dampening materials. There are two types of surface-damping treatments for body panels. Unrestrained or free-layer damping treatments, which just use the base layer and damping layer, are the initial type of damping treatment, as shown in Figure 9.1.

The second type of surface damping treatment is confined or multi-layer, which includes more than two layers. In this example, a viscoelastic-damping layer is put to a base layer and is secured by a third layer. These designs are frequently used in partly constrained layer (PCL) damping treatments to produce extensional and shear damping. This section treats these two different kinds of composite material damping treatments as two-DOF and three-DOF systems, respectively. The results are shown in various graphs to compare the damping treatment between two-layer and three-layer composite materials. Conventional techniques like experimental methods or laser vibrometry are frequently used to optimize design aspects such as material type, size, and damping treatment site. In order to locate flexible zones on structural panels, the structure must be triggered for a broad frequency range at all noise transmission paths. This makes experimental techniques cumbersome, time-consuming, and prohibitively expensive. To get around the constraints of experimental procedures, engineers must find efficient analytical techniques that can be used early in the vehicle design process to determine the appropriate damping treatments. The modal analysis of a sectional panel provides designers with an initial estimation of system properties, such as damping and stiffness coefficients (Figure 9.2).

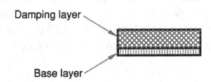

FIGURE 9.1 Damping treatment of two layers.

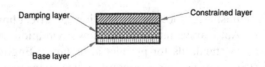

FIGURE 9.2 Damping treatment of three layers.

9.3 CASE STUDIES

9.3.1 MODAL ANALYSIS

The hood panel, sometimes referred to as the engine cover or bonnet, for passenger cars, is a significant example of the SMC technology being employed in construction. A hood panel typically has two layers, which are the outer and inner panels. SMC makes the hood panels for the Ford, Lincoln, Continental, Chrysler, Sebring JX, Dodge Viper, GM Corvette, Camaro, EV1, and other automobiles. The two fabrication formulas used in this case study to describe the outer and inner panels of a conventional car hood were developed and introduced by the Hyundai Motor Company. Low-density hollow glass micro-sphere-filled LPMC is used to construct the inner panel, whereas general-density low-profile LPMC is used to construct the outer panel. The outer and inner panels are manufactured in zinc alloy molds and then connected at room temperature with an adhesive. Study [1] lists the changes Hyundai made to the LPMC compound for the hood's outer and inner panels. It should be noted that the resin contains styrene monomer, ortho-type UP, iso-type UP, crystalline saturated polyester, and ortho-type UP. Comprehensive explanations of materials, compounding, and molding can be found in the technical papers [1,2].

The tensile, flexural, and impact properties of LPMC materials may be evaluated via a universal testing machine (UTM) in accordance with the procedures described in ASTM D638, D790, and D256, respectively. A summary of the Hyundai hood panels' material properties may be found in study [1]. The impact of the road on the body frame or chassis of a moving car causes vibrations that are transmitted to the hood panel, causing noise and discomfort. When determining the stiffness and damping coefficients of the latch/hook assembly, hinges, rubber stop, and hood panel itself, designers must take the hood's reaction into account. A simple model for the LPMC composite material bonnet of a passenger car is suggested in this section. Following parameter estimations for the system, designers perform a vibration analysis and generate differential equations of motion as a first step in determining the overall response of the hood and conducting further research. Based on the response, the system settings can be created as necessary.

According to study [1], the hood panel is fastened to the body from the front by a latch and hook mechanism, and from the back by hinges. On the side of contact between the hood and the body, the rubber-stop components are also used to reduce noise or isolate vibration. Our lumped model, which is depicted in study [1], represents a car's four wheels as a single wheel. The car's engine and bonnet are attached to the chassis directly in this configuration. Additionally, shock absorbers connect the frame to the wheels. In Table 9.1, which stands for mass, stiffness, and damping coefficients, respectively, descriptions of the letters m, k, and c are provided.

TABLE 9.1
Design Parameters

Design Parameter	Description
m_1	Effective mass of carbon fiber hood
m_2	Effective mass of engine block
m_3	Effective mass of body frame or chassis
k_1	Effective stiffness of hinges and latch/hook between hood and frame + structural stiffness of composite hood
k_2	Effective stiffness of mount rubber between engine block and frame
k_3	Effective stiffness constant of shock absorbers
c_1	Effective damping coefficient of rubber-stops + structural damping of composite hood
c_2	Effective damping coefficient of engine mount rubber
c_3	Effective damping coefficient of shock absorbers

It should be noted that the rubber stops, hinges, and latches, which are connected to the composite hood's structural rigidity in series, all have stiffness coefficients, are added in parallel to form the effective stiffness, or k_1. The stiffness of rubber stops can be ignored when contrasted to the high stiffness coefficients of hinges and latch/hook assemblies. The effective damping coefficient, or c1, is the parallel sum of the damping coefficients of the hinges, latch, and rubber stops, just like the structural damping of the composite hood. The impact of damping on hinges and latch/hook assemblies can also be ignored, in contrast to the higher damping coefficients of rubber stops. The automobile is supposed to experience an input displacement variation of $Y = Y_0 \sin(\omega t)$, which reflects a sinusoidal model of road profile, as seen in study [1]. With a wavelength of $L = 6$ m, it is safe to assume that the maximum amplitude of this displacement will be close to 0.05 m. The vehicle's speed can also be calculated as $V = 40$ mph or 18 m/s.

Finally, the system is assessed with the base wheel rotating harmonically. For different vehicle speeds, the corresponding solutions can be plotted. A free-body diagram and Newton's second law are used to find the model's governing equations of motion. $\sum F = m_i \ddot{X}_i$, where $i = 1$, 2, and 3. Figure 9.3 shows the free-body diagram of forces; using Newton's rule, the matrix form of the equations of motion may be stated as:

$$[m]\ddot{X} + [c]\dot{X} + [k]X = Y \tag{9.1}$$

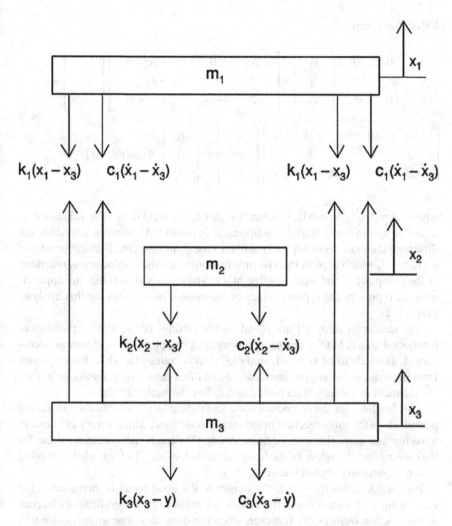

FIGURE 9.3 Free body diagram of the vehicle's hood.

These equations can be expanded as:

$$
\begin{bmatrix} m_1 & 0 & 0 \\ 0 & m_2 & 0 \\ 0 & 0 & m_3 \end{bmatrix} \begin{Bmatrix} \ddot{x}_1 \\ \ddot{x}_2 \\ \ddot{x}_3 \end{Bmatrix} + \begin{bmatrix} 2c_1 & 0 & -2c_1 \\ 0 & c_2 & -c_2 \\ -2c_1 & -c_2 & 2c_1+c_2+c_3 \end{bmatrix} \begin{Bmatrix} \dot{x}_1 \\ \dot{x}_2 \\ \dot{x}_3 \end{Bmatrix} +
$$

$$
\begin{bmatrix} 2k_1 & 0 & -2k_1 \\ 0 & k_2 & -k_2 \\ -2k_1 & -k_2 & 2k_1+k_2+k_3 \end{bmatrix} \begin{Bmatrix} x_1 \\ x_2 \\ x_3 \end{Bmatrix} = \begin{bmatrix} 0 & 0 & 0 \\ 0 & 0 & 0 \\ 0 & 0 & c_3 \end{bmatrix} \begin{Bmatrix} 0 \\ 0 \\ \dot{y} \end{Bmatrix} + \begin{bmatrix} 0 & 0 & 0 \\ 0 & 0 & 0 \\ 0 & 0 & k_3 \end{bmatrix} \begin{Bmatrix} 0 \\ 0 \\ y \end{Bmatrix}
$$

$$(9.2)$$

Using $Y = Y_0 \sin(\omega t)$:

$$
\begin{bmatrix} m_1 & 0 & 0 \\ 0 & m_2 & 0 \\ 0 & 0 & m_3 \end{bmatrix} \begin{Bmatrix} \ddot{x}_1 \\ \ddot{x}_2 \\ \ddot{x}_3 \end{Bmatrix} + \begin{bmatrix} 2c_1 & 0 & -2c_1 \\ 0 & c_2 & -c_2 \\ -2c_1 & -c_2 & 2c_1+c_2+c_3 \end{bmatrix} \begin{Bmatrix} \dot{x}_1 \\ \dot{x}_2 \\ \dot{x}_3 \end{Bmatrix} +
$$

$$
\begin{bmatrix} 2k_1 & 0 & -2k_1 \\ 0 & k_2 & -k_2 \\ -2k_1 & -k_2 & 2k_1+k_2+k_3 \end{bmatrix} \begin{Bmatrix} x_1 \\ x_2 \\ x_3 \end{Bmatrix} = \begin{Bmatrix} 0 \\ 0 \\ 1 \end{Bmatrix} A \sin(\omega t - \alpha) \tag{9.3}
$$

where, $A = Y_0\sqrt{k_3^2 + (c_3\omega)^2}$, $\alpha = \tan^{-1}(-c_3\omega / k_3)$, $\omega = 2\pi V/L$ are the values for A, α and ω. Given some initial conditions, it is possible to numerically solve the differential equations in Eq. (9.3) to obtain the solutions. One illustration of this is study [1], which depicts the system's response and the displacement reactions of the composite hood panel, engine block, and chassis. The following approximations represent the typical values of the system parameters for this analysis (Table 9.2).

The results in study [3] are based on the settings of the system parameters mentioned above, but the designer can engineer or optimize the results as necessary. This needs to be stressed, study [3] demonstrates that while the responses from the mass of the engine are separate and often have a bigger amplitude, they are identical in a steady-state period and follow the same pattern.

This implies that the composite hood and the car structure, after a transitional period, show the same reaction in the steady-state zone, which is very effective in lowering and managing overall noise. Study [3] depicts the system response for the case where the weight of the hood is doubled, or $m_1 = 2 \times 9$ kg, while the other system parameters are left the same.

Figure 9.3 shows that when the weight of the hood panel is increased to the usual weight of a steel hood, there does not seem to be a significant difference in the reaction ($m_1 = 18$ kg), however, when the transitory time grows, noise will also grow. Additionally, the weight of the hood affects the vehicle's aerodynamics even if the composite hood reduces weight and improves fuel efficiency. As seen in Figures 9.4 and 9.5, the hood's reaction pattern closely mirrors that of the frame. Rubber stops that are used as vibration isolators should be set up to have a

---

TABLE 9.2
Typical Values of System Parameters

Parameters	Parameters	Parameters
$m_1 = 9$ kg (20 lb)	$m_2 = 272$ kg (600 lb)	$m_3 = 136$ kg (300 lb)
$k_1 = 22 \times 10^4$ N/m	$k_2 = 20 \times 10^4$ N/m	$k_3 = 3 \times 10^4$ N/m
$c_1 = 85$ N·s/m	$c_2 = 8851$ N·s/m	$c_3 = 1.15 \times 10^4$ N·s/m

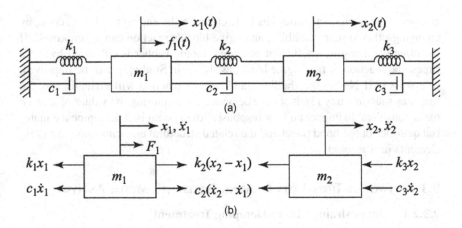

FIGURE 9.4 (a) Two DOF model and (b) FBD.

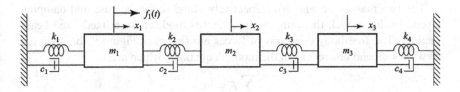

FIGURE 9.5 DOF model.

reduced response if vibration is a problem for the car hood. It would be interesting to compare the results if the suggested model's effective damping coefficient for the composite hood panel increases. Study [3] shows the response in the case of a higher effective damping coefficient, such as $c_1 = 10.85$ N·s/m. This illustration illustrates that the composite hood's reaction is over-damped with less overshoot than in the preceding case. Due to the brief duration of the transitory time, the hood panel quickly approaches having the same reaction as the body structure. This encourages the hood to vibrate locally as little as possible, hence reducing noise. However, as seen in study [3], both the amplitude and transient duration rise if the effective damping coefficient decreases by an order of 10, $c_1 = 8.5$ N·s/m. When compared to the earlier situations in Figure 9.5, this could make more noise and be more uncomfortable.

The number of vibration cycles in the transient region will decrease if the effective stiffness coefficient of the composite hood panel, k_1, drops by an order of 10, to $k_1 = 22 \times 10^3$ N/m, as shown in study [3]. In the steady-state zone, however, there will be a phase shift between the input and hood response. Study [3] shows that when the stiffness coefficient is increased by an order of 10, the responsiveness of the hood panel improves in the steady-state region even when noise is present in the transient state, making $k_1 = 22 \times 10^5$ N/m. Although altering the associated composite hood characteristics has little effect on the engine block's reaction, by changing other system variables, an engine block's reaction can be improved. If the effective damping coefficient of engine mount rubber is enhanced by three times, the reaction of the engine block, as shown in Study [3], will be improved, or $c_2 = 3 \times 8851$ N·s/m, But the hood and frame's reaction will virtually stay the same, as seen in study [3]. It should be noticed that altering the values of k and c has a considerable impact on how responsive the system is. The composite material qualities of the hood panel and the related structural elements may take these elements into account.

9.3.2 TWO- OR THREE-LAYER DAMPING TREATMENT'S MODAL ANALYSIS

9.3.2.1 Unconstrained Layer Damping Treatment

The two-layer unconstrained damping technique shown in Figure 9.1 may be represented by the damped two degree of freedom (DOF) system shown schematically in Figure 9.4a.

The two masses, m_1 and m_2, respectively, stand in for the base and damping layers. In this model, the damping treatment is modeled by a fixed-fixed beam method. The free-body diagrams of forces are shown in Figure 9.4b. Applying Newton's second law results in the motion equations, which are:

$$\sum F = m_i \ddot{X}_i$$

where, $i = 1, 2$

$$m_1 \ddot{x}_1 + (c_1 + c_2)\dot{x}_1 - c_2 \dot{x}_2 + (k_1 + k_2)x_1 - k_2 x_2 = f_1(t) \tag{9.4}$$

$$m_2 \ddot{x}_2 + (c_2 + c_3)\dot{x}_2 - c_2 \dot{x}_1 + (k_2 + k_3)x_2 - k_2 x_1 = 0 \tag{9.5}$$

Which in matrix form is:

$$[m]\left\{ \begin{array}{c} \ddot{x}_1 \\ \ddot{x}_2 \end{array} \right\} + [c]\left\{ \begin{array}{c} \dot{x}_1 \\ \dot{x}_2 \end{array} \right\} + [k]\left\{ \begin{array}{c} x_1 \\ x_2 \end{array} \right\} = \left\{ \begin{array}{c} f_1(t) \\ 0 \end{array} \right\} \tag{9.6}$$

Mass, stiffness, and damping matrices are:

$$m = \begin{bmatrix} m_1 & 0 \\ 0 & m_2 \end{bmatrix}, \quad [c] = \begin{bmatrix} c_1 + c_2 & -c_2 \\ -c_2 & c_2 + c_3 \end{bmatrix}, \quad [k] = \begin{bmatrix} k_1 + k_2 & -k_2 \\ -k_2 & k_2 + k_3 \end{bmatrix}$$

Numerical techniques are recommended to solve the set of differential equations in Eq. (9.6) because continuing with analytical procedures is often time-consuming. On the other hand, with low DOF, the impedance approach can be utilized to solve the issue and derive the solutions analytically. The impedance approach views responses and force as complex variables, which is $x_j(t) = X_j e^{i\omega t}$ where $j = 1$, 2 and $f_1(t) = F_1 e^{i\omega t}$, which is later substituted in Eq. (9.4) which gives us:

$$\left[-\omega^2 m_1 + i\omega(c_1 + c_2) + k_1 + k_2\right]X_1 - (i\omega c_2 + k_2)X_2 = f_1(t) \qquad (9.7)$$

$$\left[-\omega^2 m_2 + i\omega(c_2 + c_3) + k_2 + k_3\right]X_2 - (i\omega c_2 + k_2)X_1 = 0 \qquad (9.8)$$

Equations (9.7) and (9.8) for x_1 and x_2 must be calculated simultaneously. To obtain accurate results, it is required to obtain the true quantity of complex variables. The actual responses can therefore be summarized as follows:

$$x_j(t) = \text{Re}\left(X_j e^{i\omega t}\right) = \text{Re}\left(X_j \cos \omega t + iX_j \sin \omega t\right) \qquad (9.9)$$

As previously said, analytical approaches for low DOF systems, such as impedance and Laplace methods, are appropriate; however, numerical methods are more promising, especially for high DOF systems.

9.3.3 CONSTRAINED LAYER DAMPING TREATMENT

The three-layer restricted damping approach shown in study [3] can be replicated using three DOF systems (see Figure 9.5). Using a free-body diagram of the system and Newton's second law, the differential equations of motion for this mode are found and provided:

$$m_1 \ddot{x}_1 = -k_1 x_1 + k_2(x_2 - x_1) - c_1 \dot{x}_1 + c_2(\dot{x}_2 - \dot{x}_1) + f_1(t) \qquad (9.10)$$

$$m_2 \ddot{x}_2 = -k_2(x_2 - x_1) - k_3(x_2 - x_3) - c_2(\dot{x}_2 - \dot{x}_1) - c_3(\dot{x}_2 - \dot{x}_3) \qquad (9.11)$$

$$m_3 \ddot{x}_3 = k_3(x_2 - x_3) - k_4 x_3 + c_3(\dot{x}_2 - \dot{x}_3) - c_4 \dot{x}_3 \qquad (9.12)$$

Re-arranging the above equations we get:

$$[m]\begin{Bmatrix} \ddot{x}_1 \\ \ddot{x}_2 \\ \ddot{x}_3 \end{Bmatrix} + [c]\begin{Bmatrix} \dot{x}_1 \\ \dot{x}_2 \\ \dot{x}_3 \end{Bmatrix} + [k]\begin{Bmatrix} x_1 \\ x_2 \\ x_3 \end{Bmatrix} = \begin{Bmatrix} f_1(t) \\ 0 \\ 0 \end{Bmatrix} \qquad (9.13)$$

where mass, stiffness, and damping matrices are:

$$[m] = \begin{bmatrix} m_1 & 0 & 0 \\ 0 & m_2 & 0 \\ 0 & 0 & m_3 \end{bmatrix}, \quad [c] = \begin{bmatrix} c_1 + c_2 & -c_2 & 0 \\ -c_2 & c_2 + c_3 & -c_3 \\ 0 & -c_3 & c_3 + c_4 \end{bmatrix},$$

$$[k] = \begin{bmatrix} k_1 + k_2 & -k_2 & 0 \\ -k_2 & k_2 + k_3 & -k_3 \\ 0 & -k_3 & k_3 + k_4 \end{bmatrix}$$

In this chapter, the numerical method based on Runge-Kutta methods is used to solve the set of differential equations in Eq. (9.13) analytically, similar to the impedance approach. The parameters used in this analysis are as described in study [3].

Study [3] represents the free vibration and forced vibration responses of a two-layer composite material for an unconstrained layer damping treatment, respectively. Free vibration happens when $f_1(t)=0$. Take note of this. It is assumed for the forced vibration that the car is traveling at a speed of $V=40$ mph (or 18 m/s) along a rough road. Furthermore, it is thought that the excitation force is harmonic to the 20 Hz excitation frequency. The study shows that the damping treatment causes the response to diminish with time, as shown in study [3]. Without the second layer, the action of the springs $c_2 \approx c_3 \approx 0$ will be dominating and the response of the panel won't be dampened. The reaction will gradually worsen over time, assuming very little structural damping.

The effectiveness of damping treatment is better evaluated using forced vibration analysis when there is external stimulation on a two-layer panel. The forced vibration is brought on by the vehicle's motion as it moves along the road. As shown in study [3], the second layer's reaction is much less than the base layer's in both transient- and steady-state responses. This difference is transmitted to the other components and the passenger compartment. By doing this, background noise is reduced and the system's tolerance for fatigue is increased. For instance, the reaction amplitude of the second layer is now 50% lower than that of the first layer. This is seen in study [3], which shows that the results of a restricted layer damping treatment show the free vibration and forced vibration responses of a three-layer composite material. Free vibration analysis enhances the third layer's reaction, which is transmitted to the other parts and the passenger area, as shown in study [3]. Despite having a somewhat longer decay period, the overshoot and overall amplitude of the response are lower than those of a two-layer treatment when compared to the results shown in study [3].

Additionally, the third layer's responsiveness has greatly increased in both transient and steady-state modes, as seen by the results of the forced vibration investigation shown in study [3]. In steady-state mode, the response's amplitude is roughly 40% lower than that of the base layer.

The response amplitudes are successfully reduced in comparison to the results of the unconstrained damping treatment for forced vibration in study [3], despite a little increase in transient time.

These assessments show that damping treatments are frequently effective in reducing and controlling the noise and vibration of body panels, which enhances fatigue resistance. With the three-layer limited damping treatment, we get better results than with the unconstrained damping treatment, but there is always a cost trade-off.

9.4   CONCLUSION

The market for high-performance cars currently nearly exclusively uses composite materials. Two unique case studies addressing the vibration analysis of modern, common composite materials used in car body panels are discussed in this chapter.

The main goal of this chapter is to present readers with a technique for modeling, assessing, and controlling vibration and noise in any field of engineering using examples from the body of a car. Standard metal structural panels with low structural damping result in a system with nearly undampened forced vibration; the overall response will cause sizeable oscillations that may result in resonance for the specific body panel. However, it should be noted that if the structural panels are constructed from a suitable composite material with a good damping coefficient or damping treatment, the noise and vibration may become manageable and hence lessen passenger discomfort as well as structural fatigue.

REFERENCES

[1] I. Y. Shen, and P. G. Reinhall, "Surface damping treatments: Innovation, design, and analysis," Technical Report, 17 June, Seattle, University of Washington, 2001.

[2] C. H. Choi, S. S. Park, K. W. Ahn, and J. E. Rhee, "Low pressure molding compound hood panel for a passenger car," *FISITA World Automotive Congress*, Seoul, Korea, 2000.

[3] A. Elmarakbi, *Advanced Composite Materials for Automotive Applications*, 2013, John Wiley & Sons Ltd, United Kingdom.

10 Additive Manufacturing in Composites
Fundamentals of Processes

10.1 WHAT IS AM?

One of the most important advances of mankind was the introduction of digital technology and its replacement instead of analog processes. Thanks to this revolution in technology, many transformations have been achieved in the field of communication, imaging, architecture, and engineering in recent decades. The position of the additive manufacturing process in the manufacturing world can be considered like the analog process in the digital world. In fact, the presence of additive manufacturing has created a great transformation in the world of parts manufacturing and production. Thanks to this technology, the flexibility and efficiency that the digital system has provided for the world of electronics were provided for the world of construction and production.

Additive manufacturing is the official name of processes such as rapid prototyping and 3D printing. The word "Rapid Prototyping" or "RP" refers to a set of industrial processes, thanks to which the system or part of its components are produced before the final sample is produced. In other words, in these production methods, the emphasis is on the high speed of production and output in the form of a basic model or prototype. Therefore, the prototype produced will be different from the final sample in terms of quality. Although the combination of "rapid prototyping" has a different concept from the point of view of software engineers,[1] but in the world of production, the meaning of rapid prototyping refers to the technology in which the physical model is produced directly from the digital model data. Although this technology was initially only used for making prototypes, today it is used in making final parts (and not prototypes) for different purposes.

According to the recent widespread applications, the users of RP technology realized that the title of rapid prototyping does not cover the extent of the progress of its applications in recent years. Considering that the higher the output quality of these machines is, the quality of the final product also increases, as a result, attention was focused on upgrading these machines. After some efforts in this field, researchers and manufacturers decided to basically produce the final product directly with this method. Therefore, the word "prototyping" could no longer be used in the title of this process, because in this process, not only the prototype

DOI: 10.1201/9781003429197-10

or part but all the components were done with an additive approach (adding material step by step). Therefore, the committee formed by ASTM decided to adopt a new name for this process.[2] Although this issue is still a matter of debate in some forums, the ASTM standard forum called this process "additive manufacturing."

In additive manufacturing, the data, codes, and commands of computer-aided design software (CAD) or 3D scanners are used, and the hardware can inject materials layer by layer with precise geometry based on computer data on the workpiece. As the name implies, in additive manufacturing, materials are injected with the purpose of making a specific part. On the other hand, when a part is made in a traditional way, often postprocessing processes such as milling and machining are needed to remove extra components.

The basic principle of additive manufacturing, or AM, is to first produce a model by 3D computer-aided design (3D CAD). In the AM process, this model can be built directly without the need for planning. However, this process is not as simple as what was stated, but in fact, the production of the product using the additive manufacturing method is much simpler than other production methods. In other production methods, according to the desired dimensional tolerance, it is sometimes necessary to use several separate machines to make the part with the desired precision. In contrast, AM only requires some basic dimensional details, mastery of how the AM machine works, and knowledge of materials for fabrication.

The key point in how AM works is that the final product is created by adding material in layers, where each layer is a small cross-section and thickness of the original product, the order of which is determined by the CAD data. It is obvious that each layer must have a specific thickness so that the resulting part can be produced with a good approximation to the model. In parts (a)–(d) of Figure 10.1, respectively, the CAD model, faced model, virtual cut model, tool movement path, layers created on top of each other, and finally the made cup are shown.

According to Figure 10.1, the smaller the thickness of the layers, the closer the manufactured part will be to the software model with proper accuracy. It should be noted that in Figure 10.1, the partial production steps are presented using the additive manufacturing method. The map of the additive manufacturing process along with the main parts of the product production is shown in Figure 10.2.

Although additive manufacturing may seem new to some, the working life of this technology goes back several decades. Thanks to the remarkable performance of additive manufacturing, it is possible to manufacture parts with complex geometry while keeping the process simple. As a result, the opportunity for activists in the field of additive manufacturing is very wide. According to some, additive manufacturing is the first manufacturing technology in the world that allows engineers to improve product performance and high reliability while providing the highest production speed. For further understanding, first the applications of additive manufacturing over other methods will be presented in the next section.

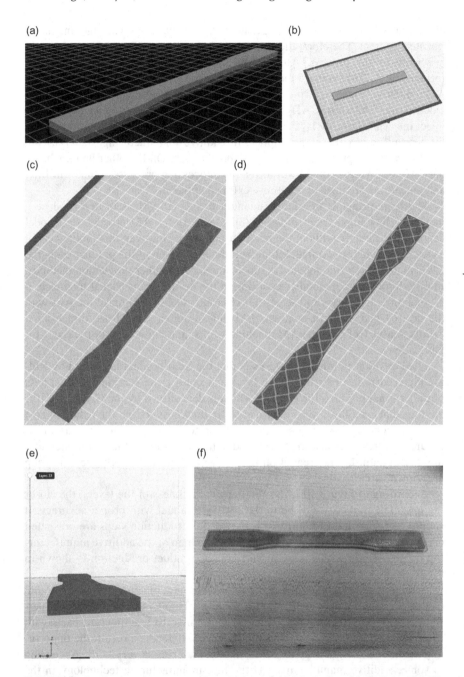

FIGURE 10.1 Systematic process of producing by AM including (a) CAD model, (b) faced model, (c) model with virtual cutting, (d) tool movement path, (e) layers created on top of each other and (f) manufactured part.

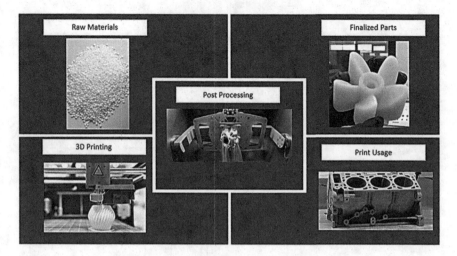

FIGURE 10.2   The process map of the main parts of product production using the additive manufacturing method.

10.2   WHAT ARE TYPES OF SHAPING?

According to ISO 52900-2015, shaping of materials into specific objects within a manufacturing process can be achieved by one, or combinations of three basic principles as follows including (1) formative, (2) subtractive, and (3) additive shaping as shown in parts (a), (b), and (c) of Figure 10.3, respectively.

In formative shaping, the desired shape is produced by the application of pressure to a body of raw material. Forging, bending, casting, injection molding, and stamping are some examples of formative shaping [1]. This type of shaping is best suitable for high-volume production for a specific part, which needs a large initial investment in tooling at first and then being able to acquire parts quickly with a low unit price [2].

Subtractive shaping is a shaping process in which the desired shape is produced by selective removal of material. Some subtractive shaping processes are milling, turning, drilling, and EDM [1]. It is believed that subtractive manufacturing, which lies between formative and additive, is suitable for parts with relatively simple geometries and produced at low-mid volumes [2].

In additive shaping process, the desired shape is produced by the addition of material. This type of process is suitable for low-volume parts with complex geometries and when a unique rapid shaping is needed [2].

Since cost is the governing factor to decide how a part will be produced and to compare formative, subtractive, and additive manufacturing shaping, these three shaping processes are compared in terms of cost per part and the number of parts in Figure 10.4. It can be concluded that the cost per part varies based

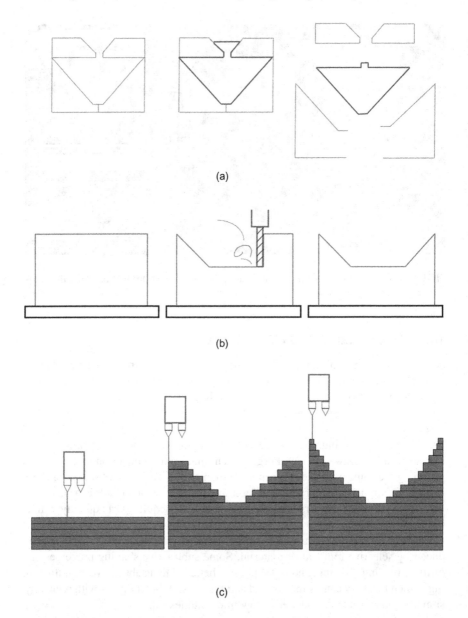

FIGURE 10.3 Types of shaping; (a) formative, (b) subtractive, and (c) additive.

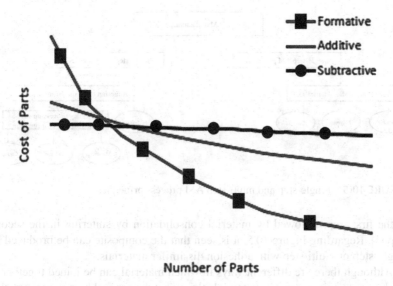

FIGURE 10.4 Comparison of economies of scale of formative, subtractive, and additive shaping.

on the amount of parts being manufactured. Also, although for high-volume production formative shaping is applicable, for low-volume production additive shaping is the best option. Regarding this, the following different types of additive manufacturing process will be presented.

10.3 WHAT ARE TYPES OF ADDITIVE MANUFACTURING PROCESS?

In additive manufacturing process, the products' fundamental properties can be determined by [1]:

1. Types of material including ceramic, metal, polymer, or composite.
2. Principle applied for fusion or bonding including sintering, curing, and melting.
3. Feedstock that is used for adding materials including powder, liquid, filament suspension, and sheet.

Regarding this, the AM can be categorized into different types. In another point of view, additive manufacturing can be divided into single-step AM and multi-step process as shown in Figure 10.5.

Forming three-dimensional parts by the addition of material is the basic principle of additive manufacturing. As shown in Figure 10.5, the basic geometry and final characteristics will be provided in a single process step (single-step AM process) or manufacture the geometry in a primary process step and ten manufacture the final characteristics of part (multi-step AM process). For instance, the desired part or object manufactures the basic geometry by joining material with a binder

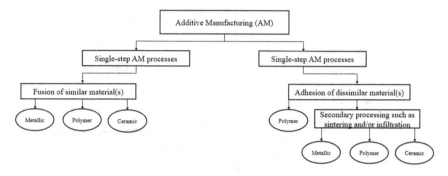

FIGURE 10.5   Single-step and multi-step AM process principles.

in the first step, followed by material consolidation by sintering in the second step [1]. Regarding Figure 10.5, it is seen that the composite can be produced by single-step or multi-step with adhesion dissimilar materials.

Although there are different ways in which material can be joined together to produce a part, in composite materials the object is acquired by any combination of metallic bonds (for metallic material), covalent bonds (for polymer materials), and ionic or covalent bonds (for ceramic materials). Thus, it can be concluded that different types of additive manufacturing process in composites can be achieved. In the following sections, some of the additive manufacturing processes including extrusion-based, powder-based, photopolymerization-based, and alternative and hybrid process will be introduced.

10.4   EXTRUSION-BASED PROCESS

Extrusion-based AM is a method to construct 3D part layer by layer using different types of feedstock materials such as viscoelastic inks of thermosets, highly viscous hydrogel, and thermoplastic filaments and pellets. In this process, extruders, syringes, and other printheads are used to deliver materials with external pressures. This external pressure can be produced by piezoelectrical stimulation, mechanical force, and pneumatic pressure. Most of the extrusion-based processes can be applied to manufacture short-fiber reinforced polymeric composite. In this process, continuous fibers are blended with thermoset precursors or thermoplastics. Fused filament fabrication (FFF) and direct ink writing (DIW) are two types of extrusion-based processes [3].

10.4.1   Fused Filament Fabrication (FFF)

Fused filament fabrication (FFF) is a type of material extrusion AM in which thermoplastic filament is employed to feed into the heated printing nozzle and extrude the molten material onto a bed. This process can be achieved by pre-programmed printing path in a layer-by-layer process. The schematic of the process is shown in study [4].

As shown in study [4], in the FFF process thermoplastic polymer filament melts inside a print head. Then, molten material extrudes onto a print bed. This process will be proceeded layer by layer. Note that the print head extrudes materials for a single *XY* plane and moves *Z* axis by the thickness of one layer and repeats until the final object is produced. Acrylonitrile butadiene styrene, polylactic (PLA), ABS, polycarbonate (PC), acrylonitrile styrene acrylate, glycol-modified polyethylene terephthalate, high-impact polystyrene, and polyamide (nylon) are some of materials, which can be used in FFF process [4]. The schematic of the structure of fiber-reinforced composite part is shown in study [4]. As shown in study [4], defects including voids exist in the filament. Then, this turns into voids in the printed beads. These voids will act as stress concentration in the object and reduce its performance. To avoid this, some approaches including infrared heating of a deposited layer are introduced [5].

10.4.2 Direct Ink Writing (DIW)

Direct ink writing or DIW is one of the newly developed methods of extrusion-based AM process. In this process, additives such as fibers are added to resin (which is thermoset and called "inks") to modify the viscosity of feedstock materials. Thus, these become sufficiently viscous to hold shape during the AM process [3], and there is no need for a pre-formed filament, which is widely used in extrusion-based AM process [6]. The schematic of the process is shown in part (a) of Figure 10.6.

Polymer matrix composites (PMC), which are composite materials with a polymer matrix and fiber reinforcements or powder, can be manufactured by DIW process. In part (b) of Figure 10.6, the Mitsubishi electric Melta RV-2F robotic manipulator, which is used for producing electrically conductive PMCs

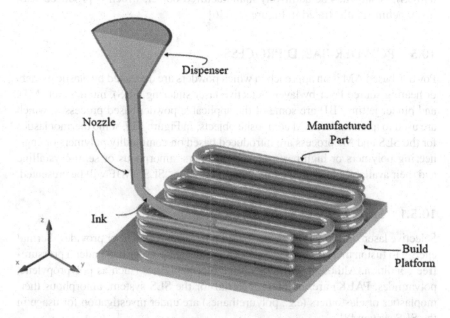

FIGURE 10.6 The direct ink writing (DIW) process, which includes additively manufactured object.

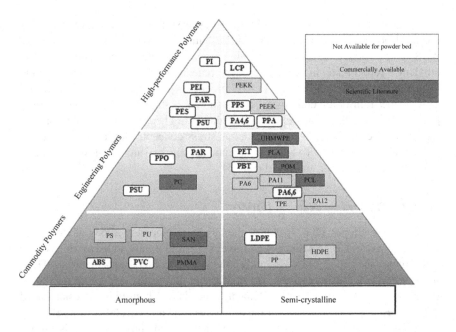

FIGURE 10.7 The thermoplastics for SLS or MJ processes including amorphous, semi-crystalline, and their availability.

(eMCS), is shown. The additively manufactured object, which is produced with this machine is illustrated in Figure 10.6 [6].

10.5 POWDER-BASED PROCESS

Powder-based AM is an approach in which powders are deposited by plastic binders or heating source layer by layer. Selective laser sintering (SLS), multiple jet (MJ), and binder jetting (BJ) are some of the applicable powder-based processes, which are used to manufacture AM composite objects. In Figure 10.7, some thermoplastics for the SLS and MJ process are introduced based on commodity polymers or engineering polymers or high-performance polymers, amorphous or semi-crystalline and their availability [7]. In the following subsection "SLS," MJF will be presented.

10.5.1 Selective Laser Sintering (SLS)

Selective laser sintering is a powder-based process in which laser provides thermal energy to fusion thermoplastic powder on pre-heated powder bed under a pressure-free condition. Although semi-crystalline thermoplastics (such as polypropylene, polyamides, PAEK) are the main material for the SLS system, amorphous thermoplastics or elastomers (e.g., polyurethanes) are under investigation for usage in the SLS system [8].

The addition of carbon nanofibers to the SLS process of polyamide-12 has been studied [9]. The diameters and lengths of carbon nanofibers were 60–150 nm and 30–100 μm, respectively. The different process parameters were considered including CNF-PA12 (chamber temperature of 164°C, bed temperature of 151°C, and laser power of 20 W), N-PA12 (chamber temperature of 166°C, bed temperature of 151°C, and laser power of 17 W), and AS-PA12 (chamber temperature of 170°C, bed temperature of 150°C, and laser power of 21 W). Photograph of the powder bed of the specimen and produced part of specimens CNF-PA12, N-PA12, and AS-PA 12 are shown in studies [4] and [9].

10.5.2 Multiple Jetting Fusion (MJF)

Multiple jetting fusion (MJF) process is one of the most developed powder-based additive manufacturing processes for producing polymetric components. In this process, microdroplets of resins are sprayed into the print bed by multiple nozzles for instant curing of the liquid materials through ultraviolet light as shown in Figure 10.8. Polyamides (e.g., PA11 and PA12) and polypropylene are two thermoplastics that are used in the MJ process [10].

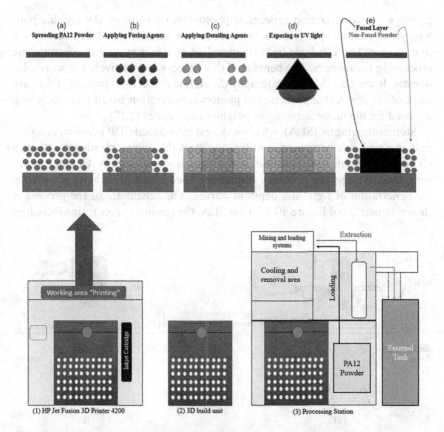

FIGURE 10.8 Schematic of multi-jet process; (1) HP Jet Fusion 3D printer 4200, (2) 3D build unit, (3) Processing station and some processing steps including; (a) Spreading PA12 powder, (b) Applying fusion agents, (c) Applying detailing agents, (d) Exposing to UV light, (e) Fused layer and non-fused powder.

The process of SLS and MJF were compared by printing polyamide 12 (PA12) in terms of physicochemical characterization of raw powder materials (EOS PA2200 and HP 3D HR PA12) [11]. To compare geometrical characteristics, the deviation degrees of both scanned objects with respect to the CAD model are provided in parts study [11] for SLS and MJF processes, respectively. The mechanical characteristics including flexural and tensile characteristics of SLS and MJF objects were also investigated.

According to parts study [11], it is concluded that MJF-produced part had lower profile deviations (1.1344–0.668 mm) and the profile deviations of SLS-produced part were up to 3 mm. The stress-strain curves of the SLS and MJF objects (study [11]) show noticeable anisotropy in tensile characteristics. Also, the tensile strength of the MJF-manufactured part was slightly higher than SLS-manufactured object.

10.6 PHOTOPOLYMERIZATION-BASED PROCESS OR VAT POLYMERIZATION (VP)

In the vat polymerization process, a photopolymer resin or UV-curable photopolymer such as epoxy and acrylic reism in a vat is utilized and cured by a light source. The UV laser light is controlled by a lens system to control some processing parameters. The benefits of this process are relatively higher resolution and lower cost. Stereolithography (SLA) and direct light process (DLP) are some of the applicable processes of photopolymerization-based process, which are used for the manufacturing of polymer composites [2,12].

Stereolithography (SLA), which was first introduced 3DP process, is a process in which a laser source is utilized to start the polymerization reaction in photocurable material. The laser movement is controlled by Galvano scanners; thus, it helps the accuracy of printing speed. The printing speed affects the penetration of light and depth of curing. The schematic of the process is shown in part (a) of Figure 10.9. In the SLA, the photopolymer resins including

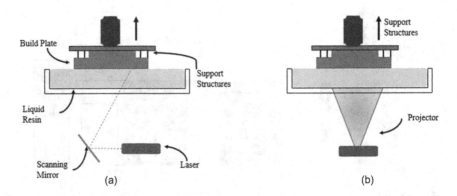

FIGURE 10.9 Photopolymerization-based process; (a) the schematic of stereolithography (SLA) process and (b) schematic of direct light process (DLP) process.

acrylate are preferred for the SLA process. Although for biomedical application (to prevent any cytotoxicity), polylactic acid (PLA) is suggested in the SLA process [12].

UV light-introduced polymerization of the resin is the fundamental of the direct or digital light process as shown in part (b) of Figure 10.9. DLP is similar to SLA, but the difference is SLA uses laser beam to cure the layers while UV light has been used to cure additive material in the DLP process [12]. Since in the SLA process laser beam can completely cure layers point by point and trace the geometry, the manufactured object has better geometry accuracy. On the contrary, the printing speed of the object in the DLP process is higher than the SLA process. In the DLP process, the micro-mirror devices (DMD) fully cure the layer at a time (which is controlled numerically).

10.7 COMPARISON OF ADDITIVE MANUFACTURING PROCESSES

AM processes, which were introduced in previous sections, can be compared in different aspects including materials, applications, benefits, drawbacks, and even resolution range as provided in Table 10.1. In terms of materials, a wide range of materials including chocolate to advanced multi-functional materials can be produced by AM process because of its noticeable advantages. As provided in Table 10.1, materials in the forms of inks, sheets, powder, wire, paste, or filament are used in the AM process. Polymers are one of the most common materials, which are used in AM in different parts in the automotive, aerospace, medical, toys, and even architectural industries. In the FDM, which is the most common method, the additive polymer filaments are used, while they are used in the form of powders and resins in the powder bed and stereolithography methods, respectively [13].

Comparing the benefits of AM process as provided in Table 10.1, it can be concluded that the FDM, direct energy deposition, and laminated object manufacturing are low-cost process, while the powder bed fusion, stereolithography, and direct energy deposition provide high-quality objects. To manufacture large structures, inkjet and laminated object manufacturing processes are suggested.

Speed, quality, and costs are three important factors that restrict the manufacturing process. In the AM process, while speed and cost are crucial, stereolithography and powder bed fusion are not recommended because they are low-speed and expensive methods. The comparison of production cost versus volume production for the AM process is provided in Figure 10.10 for further investigation. One of the disadvantages of the AM process is weak mechanical properties, which is noticeable in FDM, powder bed fusion, and direct energy deposition processes. For further investigation, in Chapter 11, the characteristics and different applications of AM process will be introduced.

TABLE 10.1

Comparison of Additive Manufacturing in Composites Methods Based on Materials, Applications, Benefits, Draw Backs, and Resolution Range

Methods	Materials	Applications	Benefits	Drawbacks	Resolution Range (μm)	References
Fused deposition modeling (FDM)	Continues filaments of thermoplastic polymers/ continuous fiber-reinforced polymers	Rapid prototyping, toys, advanced composite parts	Low cost, high speed, simplicity	Weak mechanical properties, limited materials (only thermoplastics) layer-by-layer finish	500–200	[14]
Powder bed fusion	Compacted fine powders, metals, alloys, and limited polymers (SLS or SLM) ceramic and polymer (3DP)	Biomedical, electronics, aerospace, lightweight structures (lattices), heat exchangers	Fine resolution, high quality	Slow printing, expensive, high porosity in binder method	80–250	[14]
Inkjet	A concentrated dispersion of particles in a liquid (ink or paste), ceramic, concrete, and soil	Biomedical, large structures, building	Ability to print large structures, quick printing	Maintaining workability, coarse resolution, lack of adhesion between layers, layer-by-layer finish	5–200	[15]
Stereolithography	A resin with photo-active monomers, high polymer-ceramics	Biomedical, prototyping	Fine resolution, high quality	Very limited materials, slow printing, expensive	10	[14]
Direct energy deposition	Metals and alloys in the form of powder or wire, ceramics, and polymers	Aerospace, repair, cladding, biomedical	Reduced manufacturing time and cost, excellent mechanical properties, controlled microstructure, accurate composition control, excellent for repair	Low accuracy, low surface quality, need for a dense support structure, limitation in printing complex shapes with fine details	250	[16]
Laminated object manufacturing	Polymer composites, ceramics, paper, metal-filled tapes, metal rolls	Paper manufacturing, electronics, smart structure	Reduced tooling and manufacturing time, low cost, a vast range of materials, excellent for manufacturing larger structures	Limitation and manufacturing of complex shapes	Depends on the thickness of the laminates	[13]

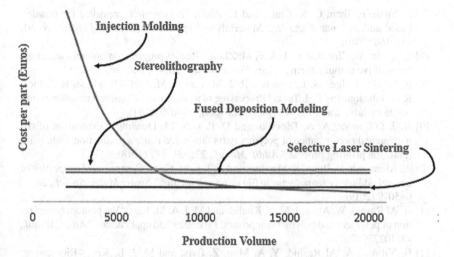

FIGURE 10.10   Cost per part for the additive manufacturing process as a function of production volume [16].

NOTES

1 Rapid prototyping is a technique in software engineering that provides useful methods for checking ideas and getting customer feedback. In this method, the initial idea (and not the final one) is presented to the customer. Then, step by step, by receiving the customer's feedback regarding the idea, the initial idea is modified and becomes the final idea with better quality.
2 F42 committee was formed in 2009 for additive manufacturing technology. This committee is held twice a year (usually spring and fall in the US and non-US respectively) with more than 150 members in 2 days. At the time of preparing this book, this committee has 725 members and 8 subgroups, all the standards related to additive manufacturing are prepared by F42 and placed in the annual book of ASTM standards.

REFERENCES

[1] B. Redwood, F. Schöffer, and B. Garret, *The 3D Printing Handbook: Technologies, Design and Applications*, 3D Hubs, 2017.
[2] S. Yuan, S. Li, J. Zhu, and Y. J. C. P. B. E. Tang, "Additive manufacturing of polymeric composites from material processing to structural design," *Compos. Part B Eng.*, 219, 108903, (2021).
[3] N. Van de Werken, H. Tekinalp, P. Khanbolouki, S. Ozcan, A. Williams, and M. J. A. M. Tehrani, "Additively manufactured carbon fiber-reinforced composites: State of the art and perspective," *Addit. Manuf.*, Netherlands, 31, 100962 (2020).
[4] V. Kishore, C. Ajinjeru, A. Nycz, B. Post, J. Lindahl, V. Kunc, and C. Duty, "Infrared preheating to improve interlayer strength of big area additive manufacturing (BAAM) components," *Addit. Manuf.*, Netherlands, 14, 7–12 (2017).
[5] S. A. Khan, and I. Lazoglu, "Development of additively manufacturable and electrically conductive graphite-polymer composites," *Prog. Addit. Manuf.*, 5(2), 153–162 (2020).

[6] S. Yuan, F. Shen, C. K. Chua, and K. Zhou, "Polymeric composites for powder-based additive manufacturing: Materials and applications," *Prog. Polym. Sci.*, 91, 141–168 (2019).

[7] L. J. Tan, W. Zhu, and K. J. A. F. M. Zhou, "Recent progress on polymer materials for additive manufacturing," *Adv. Funct. Mater.*, 30(43), 2003062 (2020).

[8] R. D. Goodridge, M. L. Shofner, R. J. M. Hague, M. McClelland, M. R. Schlea, R. B. Johnson, and C. J. Tuck, "Processing of a polyamide-12/carbon nanofibre composite by laser sintering," *Polym. Test.*, 30(1), 94–100 (2011).

[9] H. J. O'Connor, A. N. Dickson, and D. P. J. A. M. Dowling, "Evaluation of the mechanical performance of polymer parts fabricated using a production scale multi jet fusion printing process," *Addit. Manuf.*, 22, 381–387 (2018).

[10] A. Alomarah, D. Ruan, S. Masood, and Z. J. S. M. Gao, "Compressive properties of a novel additively manufactured 3D auxetic structure," *Smart Mater. Struct.*, 28(8), 085019 (2019).

[11] A. Al Rashid, W. Ahmed, M. Y. Khalid, and M. J. A. M. Koc, "Vat photopolymerization of polymers and polymer composites: Processes and applications," *Addit. Manuf.*, 47, 102279 (2021).

[12] B. Yilmaz, A. Al Rashid, Y. A. Mou, Z. Evis, and M. J. B. Koç, "Bioprinting: A review of processes, materials and applications," *Bioprinting*, 23, e00148 (2021).

[13] X. Wang, M. Jiang, Z. Zhou, J. Gou, and D. Hui, "3D printing of polymer matrix composites: A review and prospective," *Compos. Part B Eng.*, 110, 442–458 (2017).

[14] A. Kazemian, X. Yuan, E. Cochran, and B. Khoshnevis, "Cementitious materials for construction-scale 3D printing: Laboratory testing of fresh printing mixture," *Constr. Build. Mater.*, 145, 639–647 (2017).

[15] J. Edgar, and S. Tint, "Additive manufacturing technologies: 3D printing, rapid prototyping, and direct digital manufacturing," *Johnson Matthey Technol. Rev.*, 59(3), 193–198 (2015).

[16] N. Hopkinson, and P. Dicknes, "Analysis of rapid manufacturing-using layer manufacturing processes for production," *Proc. Inst. Mech. Eng. C J. Mech. Eng. Sci.*, 217(1), 31–39 (2003).

11 Additive Manufacturing in Composite
Characteristics

11.1 DESIGN FOR AM POLYMER COMPOSITES

Additive manufacturing (AM), also known as 3D printing, has revolutionized the fabrication of polymer composites, offering unprecedented design flexibility and material customization. This transformative process allows for the creation of intricate structures with tailored properties, expanding the horizons of composite materials.

11.1.1 ADDITIVE MANUFACTURING AND DESIGN FREEDOM

AM has brought a paradigm shift to the design and manufacturing of polymer composites. By layer-by-layer deposition of materials, engineers can create complex geometries and microstructures that were previously unattainable using traditional manufacturing methods. This design freedom empowers researchers to optimize material compositions and achieve desired performance characteristics.

11.1.2 THE PROCESS-STRUCTURE-PROPERTIES-PERFORMANCE (PSPP) FRAMEWORK

The PSPP framework provides a systematic approach to understand the relationship between process parameters, resulting microstructure, material properties, and overall performance of the multi-material additive manufactured polymer composite. Figure 11.1 illustrates this comprehensive framework, highlighting the interdependence of each element.

11.1.3 PROCESS

The first stage of the framework involves the AM process, where material deposition and layering take place. Key process parameters, such as printing speed, temperature, and layer thickness, influence the subsequent stages.

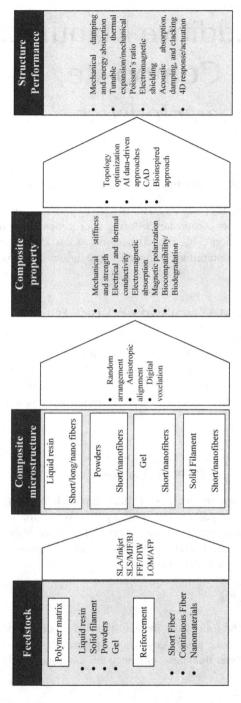

FIGURE 11.1 The process-structure-properties-performance of AM polymer composites.

11.1.4 STRUCTURE

The interaction of process parameters during AM leads to the formation of the composite's microstructure. This stage determines the arrangement and distribution of materials, reinforcing phases, and voids within the composite.

11.1.5 PROPERTIES

The microstructure directly influences the material properties of the composite. This stage considers mechanical, thermal, electrical, and other relevant properties that emerge from the specific microstructural configuration.

11.1.6 PERFORMANCE

Finally, the material properties culminate in the overall performance of the additive-manufactured polymer composite. This stage assesses the composite's ability to meet specific application requirements and its potential for advanced engineering applications.

According to Figure 11.1, in the first step, the objects are manufactured from feedstock including polymer matrix (such as liquid resin, solid filament, powders, and gel) and reinforcement (such as short fiber, continuous fiber, and nanomaterials) and composite microstructure, such as liquid resin (short/long/Nano fibers), powders (short/Nano fibers), gel (short/Nanofibers), and solid filament (short/long/Nanofibers), by different AM polymer processes (such as SLA, Inkjet, SLS, and MJF). In the second step, composite properties or characteristics including mechanical, electrical, electromagnetic absorption, magnetic polarization, and biocompatibility, metallurgical and geometrical, are provided. These characteristics lead to the structure performance of objects such as mechanical damping, energy absorption, thermal expansion, electromagnetic shielding, and acoustic absorption. Thus, the realization of the functionality of the object is needed for designing materials and structures, and in the following sections, the mechanical, electrical, electromagnetic, thermal conduction and expansion, and acoustic characteristics of AM polymer composites will be considered.

11.2 MECHANICAL CHARACTERISTICS

Composite materials are designed to provide unprecedented mechanical characteristics for different types of applications. Thanks to computational optimization-based design method, which can be applied to different design scales, the performance of designed composite structures also improves. Regarding this, some mechanical characteristics such as tensile, flexural and damping, and compression are considered in AM composites by considering some related standards as provided in Table 11.1. In the following subsections, tensile and flexural and damping of AM polymer-based composites will be introduced.

TABLE 11.1

The Standards, Which Are Related to Different Mechanical Characteristics Tests ((P) for Polymer and (C) for Composite) [1]

Rows	Tests	Related Standards
1	Tensile	ASTM D638 (P)/ISO 527-2 (P)/ASTM D3039 (C)/ISO 527-4 (C)
2	Flexural	ASTM D790 (P/C)/ISO 178 (P/C)/ASTM D7264 (C)
3	Impact	Charpy Impact (Notched) ASTM D6110, ISO 179 (P/C) Izod Impact (Notched) ASTM D256, ISO 180 (P/C) Izod Impact (Unnotched) ASTM D4812, ISO 180 (P/C)
4	Compression	ASTM D695 (P)/ISO 604 (P)/ASTM D3410 (C)/ISO 14126 (C)

11.2.1 TENSILE AND FLEXURAL

Tensile and flexural properties are critical mechanical characteristics used to evaluate the performance and reliability of additive-manufactured polymer composites. These properties directly influence the material's response to various loading conditions and play a crucial role in determining its suitability for specific applications. As additive manufacturing continues to gain momentum, extensive research has been devoted to understanding and improving the stiffness and strength of these advanced composite materials.

11.2.1.1 Tensile Characteristics

Tensile properties assess a material's response to tensile forces, such as elongation, yield strength, and ultimate tensile strength. Numerous studies have been undertaken to investigate the tensile behavior of AM polymer composites. The effects of varying printing parameters, material compositions, and fiber reinforcements have been analyzed to optimize tensile performance. Researchers have found that proper alignment of reinforcement fibers and an optimized interphase between fibers and matrix significantly enhance the tensile strength of the composite.

11.2.1.2 Flexural Characteristics

Flexural properties determine a material's ability to resist bending or torsional stresses. Studies focused on the flexural behavior of AM polymer composites have shown that fiber alignment and orientation play a crucial role in influencing flexural strength and stiffness. Researchers have explored novel designs, such as lattice structures and graded materials, to enhance flexural properties while maintaining low weight. These advancements demonstrate the potential of additive manufacturing to create lightweight yet strong and stiff composite materials.

The tensile and flexural of fused deposition modeling (FDM) manufactured object was considered in a study [2]. The Continuous Fiber Reinforced Thermoplastic Composite (CFRTPCs) specimens were manufactured from nylon filament (matrix) with a diameter of 1.75 mm with reinforced fibers such as glass, carbon, and Kevlar. The specimens were divided in two types including unreinforced and continuous reinforced nylon specimens. The average diameter of carbon, glass, and Kevlar fibers were 350, 300, and 300 μm, respectively. The densities of carbon, glass, and Kevlar fibers were 1.4, 1.5, and 1.2 g/cm^3 respectively. The SEM figures of carbon, Kevlar, and glass fibers with a magnification of 5,500 are illustrated in parts (a), (b), and (c) of Figure 11.2, respectively. The details and general view of the additive manufacturing process are shown in parts (d) and (e) of Figure 11.2, respectively. As can be seen, the MarkedForged additive manufacturing process was used for the manufacturing of CFRTPCs specimens. Since the system has two extruders and two print heads, two kinds of materials (nylon (matrix) and fiber reinforcement) can print at the same time. The hot-ended temperatures 273°C and 232°C were considered in the manufacturing of nylon and fiber layers, respectively, on a non-heated print bed in the AM process. The layers of carbon, Kevlar, and glass fibers were 0.125, 0.1, and 0.1 mm.

The ASTM D3039 and D790 methods were considered. Two orientations including flat and on-edge with different orientation thicknesses of layers (including unreinforced, types A, B, and C as shown in study [2]) were introduced for the AM polymer-based composites process.

Considering study [2], it is concluded that the layer thickness affects mechanical performance of nylon specimens. The flat orientation specimens have higher tensile and flexural strength and stiffness than on-edge orientation specimens. It is revealed that carbon fiber-reinforced composites provide the best tensile and flexural strength and Kevlar fiber-reinforced composites have the lowest mechanical performance. Besides the thickness of layers, materials of composites, and orientation, other parameters such as the fiber orientation including 0, 45°, and 60° affect flexural stress and tensile stress, which were considered in other studies [1,3].

11.2.2 Energy Absorption and Damping

Polymer composites offer unique advantages, particularly in terms of energy absorption and damping capabilities. The presence of hard and soft phases within the composite structure contributes to its exceptional performance in dissipating mechanical energy under cyclic loading. This characteristic is of paramount importance in numerous engineering applications, including automotive components, sports equipment, and protective gear. The additive manufacturing process provides a platform to tailor the arrangement of these phases, enabling the creation of composite structures with optimized energy absorption and damping properties.

11.2.2.1 Energy Absorption and Damping Mechanism

The energy absorption and damping mechanism in polymer composites are closely linked to the combination of hard and soft phases. The hard phases, typically consisting of reinforcing fibers or particles, contribute to the material's overall mechanical strength and load-carrying capacity. In contrast, the soft phases, usually comprising the polymer matrix, possess viscoelastic properties that

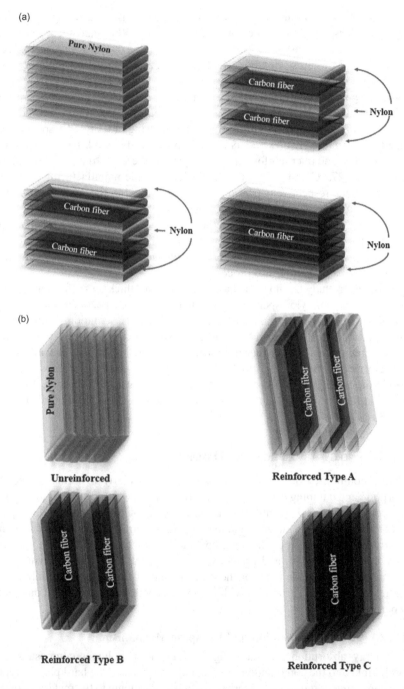

FIGURE 11.2 SEM images of fiber (5,500×), (a) carbon fiber, (b) Kevlar fiber, (c) glass fiber, (d) details of the additive manufacturing process, (e) general view of the MarkForged additive manufacturing process.

enable efficient energy dissipation during cyclic loading. This unique arrangement allows the composite to effectively absorb and dissipate mechanical energy, leading to enhanced impact resistance and vibration damping.

11.2.2.2 Tailoring Energy Absorption and Damping

Additive manufacturing offers unprecedented design freedom, allowing engineers to precisely control the spatial arrangement of hard and soft phases within the composite. By optimizing the distribution and volume fraction of these phases, the material's energy absorption and damping characteristics can be tailored to meet specific application requirements. Moreover, the use of lattice structures and gradient materials enables the creation of composites with tunable energy absorption profiles, catering to a wide range of impact scenarios [4].

Carbon Fiber Reinforced Polymer (CFRP) composite is one of the known composite structures for its high stiffness-to-weight ratio. One of the challenges of this structure is its low energy dissipation, which leads to failure [5]. A multimaterial projection microstereolithography process was considered to produce cellular materials reinforced by short fibers, which have high specific stiffness and damping coefficient [6]. As shown in study [7], a cellular material provides near-constant specific stiffness at low densities. Using a periodic architecture of unit cells also leads to desired characteristics of these structures (shown in study [7]). The lightweight CFRP microlattices including periodically arrayed unit cells with soft phase considered in out-of-plane struts were considered as shown in study [7]. It should be noted that $\tan g\ \delta = 2\Psi/\pi$, where $\Psi = \Delta U/U$ is loss efficient. ΔU is dissipated energy and U is stored energy.

To compare the performance of the stiffness-damping of proposed CFRP microlattices, specific stiffness per density ($E^{1/3}/\rho$) versus loss tangent ($\tan g\delta$) is provided in study [7].

11.2.3 COMPARISON ON MECHANICAL CHARACTERISTICS

As discussed before, different types of additive manufacturing process can be used in composites, which lead to different mechanical characteristics. To compare, tensile modulus versus tensile strength and Young's modulus versus strength are provided in studies [8,9], respectively for FDM, compression molding, stamping, injection molding, SLS, and SLA AM processes.

Regarding studies [8,9], for the case of continuous fiber-reinforced thermoplastic composite, FDM process (PLA/CF-carbon) provides above 20 GPa tensile modulus with 180–225 MPa tensile strength. The highest tensile strength is provided by compression molding (CF-glass) and stamping (CF-glass) processes. The highest strength and Young's modulus of continuous carbon-fiber composites are manufactured by CFRTP and reinforced FDM, respectively, as shown in studies [8,9]. Thus, it can be concluded that types of AM process besides additive filler material and matrix will affect the mechanical characteristics such as tensile modulus and strengths of the produced object.

11.3 ACOUSTIC CHARACTERISTICS DRIVEN DESIGN

Thanks to additive manufacturing technologies, objects with complex geometries can be manufactured. This advantage of AM process can be used to design objects, which can be used in acoustic fields including silencers or hearing protection devices [10]. Some AM technologies including FDM, FFF, SL, SLS, MJM, DLP, DMLS, which have been considered in acoustic characteristics-driven design, are provided in Figure 11.3 and Table 11.2.

Although synthetic fiber is one of the best sound absorptive materials, because of health-related concern, the researcher is studying novel alternative material such as MicroPerforated Panel (MPP). The concept of this material, which has considerable acoustic resistance with a wide range of absorption bandwidth, is based on Helmholtz resonator sound absorption. The history of proposal of MMP for sound absorption is back to the research of Professor Dah You Maa in 1975 [56]. A particular thin panel with an amount of microperforated holes is defined as an MMP [57]. Both the thickness of the panel and the diameter of microperforated holes are usually less than 1 mm, and the sum of the area of microperforated holes is between 1% and 2% of the surface area of the panel [58]. As shown in part A of study [59], an air space is introduced between a fixed wall and MPP to absorb sound. The efficiency of MPP depends on air particle motion around the microperforated holes, since the friction between air particles and the inner microperforated holes leads to converting sound energy to heat, and thus, the sound energy is absorbed.

Polylactic acid/polyhydroxyalkanoates-wood fibers (PLA/PHA-WF) composite was considered using filament with a diameter of 1.75 ± 0.05 mm with a density of 1.15 g/cm^3 in a study in fused deposition modeling additive manufacturing technology. The melting temperature and printing rate were 210°C and 70 mm/s, respectively. The thickness of MPP was 5 mm with varied infill densities including 100% (named 3DP-MPP-100), 80% (named 3DP-MPP-80), 60% (named 3DP-MPP-60), 40% (named 3DP-MPP-40), and 20% (named 3DP-MPP-20) as shown in study [59].

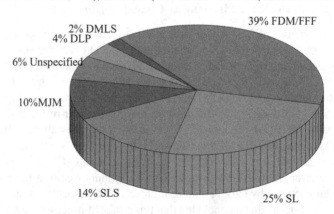

FIGURE 11.3 Some of the AM technologies considered in acoustic characteristics-driven design.

TABLE 11.2
Methods, AM Technologies and Applications for Acoustic-Driven Design Objects

Methods	AM Technologies	Application	References
Periodic cavities	FDM, SL	Low-frequency absorption	[11,12]
	FDM	Signals segregation	[13]
Labyrinth ducts	FDM, SLS	Adjustable absorption over a wide frequency range	[14–20]
Capillary ducts	FDM, SL, MJM	Acoustic absorption	[21–26]
Cellular structures	DLP, DMLS	Optimized acoustic foams	[27]
	SLS	Energy absorption	[28–30]
Modular structures	FDM	Acoustic filtration and coding	[31]
	SL	Acoustic filtration and modulation	[32]
Thin wall structures	FDM	Non-linear absorbers	[33]
	MJM	Black acoustic holes	[34–36]
Helmholtz resonators	SL	Low-frequency absorption	[37,38]
Sonic crystals	FDM, DLP	Adjustable absorption over a wide frequency range	[39–41]
	MJM, SL	Waves propagation	[42,43]
Labyrinth	FDM	Acoustic absorption	[44]
meta-surfaces	MJM	Spatial sound modulation	[45]
	FDM	Waves manipulation and modulation	[46–52]
Helical meta-surfaces	SL	Waves manipulation and modulation	[53,54]
Spatial coils	SL	Outgoing directed vortex beam	[55]

The effect of infill density on sound absorption coefficient (SAC) on MPP is shown in study [59], in which different results can be concluded. First, the maximum SAC of MPP for all specimens was nearly the same and greater than 0.9. Second, varying density by changing the mass of MPP affects slightly the resonant frequencies of the MPP. In which, by decreasing density, the resonant frequency moves toward high-frequency spectrum. Third, the reduction in density leads to performance decrease at low frequency. Thus, the density is one of the important factors, which should be considered in AM MPP process in terms of SAC.

To investigate the effect of an air gap on SAC, is studied [59], which shows there was no peak of acoustic absorption without an air gap specimen. Considering the mass-spring system (perforated end is mass and air gap is spring), without an air gap, this system will not be formed, and thus there won't exist any resonant peak (which is the natural frequency of mass-spring system). Also, the increase in air gap thickness led to a shift in the peak of sound absorption toward the low-frequency spectrum as can be seen in study [59]. Considering the perforated end as acoustic mass and air gap as acoustic spring, as the thickness of air gap increases, the stiffness decreases. Thus, the resonant frequency and resonant peak decrease and shift toward low-frequency spectrum.

Comparing study [59], it can be concluded that the maximum SAC of AM MPP is provided in the specimen with an infill density of 100% at an air gap of 15 mm with a peak of 1,100 Hz. Therefore, in additive manufacturing considering acoustic characteristics-driven design, air gap, density, and other characteristics of acoustic structure, which provided by AM affects on acoustic characteristics such as SAC.

11.4 FILLER MATERIALS IN ADDITIVE MANUFACTURING COMPOSITES

Before introducing the application of additive manufacturing in composite, in this section, some of the fillers, which are used as additives in the process, will be introduced. Additive fillers can be categorized into metallic, ceramic, carbon-based, and organic additives as shown in Table 11.3 [60].

TABLE 11.3
Polymeric Composite Fillers and Their Purposes

Filler		Matrix	Purpose
Metallic	Al	PA12	• Improved thermal and electrical conductivities and mechanical strength
	Carbon steel	PA12	
	Stainless steel	PA12, PS	
	Mo	PA12	• Metallic green part fabricated by an indirect laser sintering process
	Cu	PMMA	
			• Polymer as a removable binder
Ceramic	Al$_2$O$_3$	PA12, PEEK, Ps, PVA	• Enhance mechanical strength by micro or nano Al$_2$O$_3$ particles
	AlCuFeB	PA12	
	ZrO$_2$	PP, PA6	• Improved thermal stability
	Y$_2$O$_3$	PP	• Improved impact strength, creep resistance, toughness, and tensile strength
	SiO$_2$	PA11, PA12, PEEK	
	BaTiO$_3$	PA12	
	SiC	PA12	• BaTiO$_3$ polymeric nanocomposites with piezoelectrical properties for sensing and energy harvesting
	PTWs	PA12	
	Glass	PA11	
	Montmorillonite	PA11, PA12, PA12, PS	
	Rectorite	PA12	
	Fly ash	PEEK	
	TiO$_2$	PEEK, PA12, PET, PEN	
	CHA	PLLA, PHVB	
	HA	PEEK, PA12, PLLA, PCL, PVA, PMMA, HDPE	
	β-TCP	PLA, PCL, PLGA	
	BCP	PHVB	

(Continued)

TABLE 11.3 *(Continued)*

Polymeric Composite Fillers and Their Purposes

Filler		Matrix	Purpose
Carbon-based	CNTs	PA11, PA12	• Improved laser energy absorption and powder fusion acceleration by CNTs and graphene
	Carbon fiber	PA12, PA11, PEEK	
	Graphene	PA11	
	Graphite	PMMA, PC, PEEK	• Enhanced thermal and electrical conductivities by CNTs
	Carbon black	PA12	
			• Improved mechanical properties including specific strength, impact toughness and ductility by CNTs
			• Enhanced thermal stability and fire-retardancy by CNTs and carbon fibers
			• Light-weight composites with excellent electrical conductivity but sacrificing of mechanical toughness and tensile strength.
Organic additive	HDPE	PA12	• Improved mechanical strength and thermal stress resistance by blending PA6, PEEK or PC with PA.
	PA6	PA12	
	PS	PA12	
	PEEK	P12	• Improved processability of the composite powders with constituent PS and HDPE as binder materials at low melting temperature.
	PC	P12, PEEK	
	PTEF	PEEK	
	Epoxy	PS, PC, SMMA, PA12	• Improved processability of the composite powders with constituent PS and HDPE as binder materials at low melting temperature
	SAN	PS	
	Wax	PS, PA11	
	Brominated hydrocarbon	PA12, PA11	
	Progesterone	PCL	

Regarding Table 11.2, different results can be concluded. First, adding metallic fillers such as Al, carbon steel, stainless steel, Mo, and Cu to polymer matrix such as PA12, PS, and PMMA is considered to improve thermal and electrical conductivities and also mechanical strength in AM composites. Second, various ceramic fillers with different purposes including energy harvesting, improving thermal stability, impact strength, creep resistance, toughness, and tensile in polymer matrices such as PA11, PA12, PEEK, PS, PVA, and PMMA. Bio-ceramics CHA and HA are blended with PLLA and PCL for bioactive tissue. CNTs are famous carbon-based fillers, which can be added in the polymer matrices such as PA11 and PA12 with the aims of thermal and electrical conductivities. Adding graphene filler in PA11 matrix leads to producing lightweight composites with noticeable electrical conductivities.

Organic additive fillers such as HDPE, PA6, PS, PEEK, PC, Epoxy, SAN, and Wax will be considered in the polymer matrix (including PA11, PA12, PEEK, PC, PS, SMMA, and PCL) for different aims such as improving mechanical strength, thermal stress resistance, wear resistance or toughness. Thus, different materials can be used as additive fillers in polymer matrix during the additive manufacturing process with various aims. This leads to different applications in AM composites, which will be discussed in the next section.

11.5 GENERAL APPLICATIONS FOR AM IN COMPOSITES

The global demand for carbon fibers, which is one of the fundamental materials of composites, has increased rapidly over the last two decades in different applications and industries. In the automobile industry, carbon fiber-reinforced polymer (CFRP) occupies 17% of auto-weight, and it can reduce auto-weight by 30% [41,42]. Fifty percentage of the total weight of structural elements of aircrafts are CFRPs and thanks to it, the weight of the structural element is reduced by 20% [43]. Although it is reported that CFRP was first considered in aeronautics and astronautics in 1950s, the application of it in the civilian aerospace industry is increasing quickly [44]. The primary load-bearing structures, such as fuselages, wing planks, and sandwich panel skins, have been manufactured from composite [45]. It is mentioned that carbon fiber composites account for Boeing models 767, 777, and 7E7, with 3%, 7%, and 50% of total weight, according to Boeing's official website. The fuselage and wings of Boeing 787 Dreamliner and A350-XWB were made from composite. This can lead to a reduction of about 20% in fuel consumption (compared to airliners of the same size) [46]. In the near future, it is planned to consider composite in 50% of parts and weight of aircraft [47]. Composite also has been applied in the construction industry; the Ibach Bridge in Switzerland in 1991 was strengthened by composite [48]. To repair and improve civil structures, composite has been applied in different structures.

Additive manufacturing in composites includes various industries including automotive, aerospace, military, healthcare, biomedical engineering, water treatment and energy generation, heat exchangers, and acoustic devices. In the following sections, some of the applications of these industries will be introduced and investigated. The market size of automotive polymer composite is reported USD 6.40 billion in 2016, and it is predicted to be 11.62 billion until 2025 as shown in study [61]. It is also reported that polymer composites go for exterior components (40% of total shares) and in the next step interior components (with 25% of total shares) of polymer composites in automobile industries [61]. For further investigation, in parts studies [62,63], current and future applications of AM in the automotive industry [62] and some of the polymer composites' additive manufactured parts are shown, respectively [63].

DMLS and SLM are two AM processes that are used in the manufacturing of composite heat exchangers or HXs [64]. The first known FFFed PHX and its internal cross-section dimensions in mm are shown in study [65]. Besides the FFF process, other AM processes including vat photopolymerization, material jetting, sheet lamination, and powder bed fusion are considered in the manufacturing process of polymer composite heat exchangers [65]. Some of the features, which are considered in additively manufactured HXs are provided in study [64]. Surface roughness is one of the features, which affect directly on performance of HXs when these roughness scales are large enough to disrupt the near-wall viscous layer in turbulent flows. Microchannels are used in HXs, which are considered for small-scale heat sinks of miniaturized electronic equipment. Area and turbulence promoters directly affect the performance of HXs since increasing turbulence leads to an increase in flow, which provides higher heat performance. The complex and freeform designs are two other features that are needed for HXs and will be provided by the AM composite process.

Aerospace structures, including drone and aircraft wings, should have some special features of stiffness, strength, and stability under certain loading conditions [66]. Morphing structure as shown in study [66] can satisfy and fulfil these needs. The first composite additively manufactured morphing by using continuous carbon fiber has 2 m wingspan. A cost calculation of continuous fiber additive-manufactured morphing wing parts reports a reduction of ten times. Thus, using additively manufactured composite morphing wing is economical besides providing special mechanical features. Further case studies will be discussed in the next chapter.

11.6   ELECTRICAL AND ELECTROMAGNETIC CHARACTERISTICS

Polymer-based composite plays an important role in different applications, especially civil industries. The electrical characteristics are crucial for the electrical behavior of structure such as electromagnetic shielding ability and lightning strikes [67]. Thus, in this section, the electrical conduction and electromagnetic shielding of additive-manufactured composites will be discussed in separate subsections as follows:

11.6.1   ELECTRICAL CONDUCTION

Electrical conduction, which is the electrical characteristic of composites, is influenced by filler's distribution, the material formulation, and interfacial adhesion of filler and matrix in composites. The conductive phases in composites include metallic particles, carbon-based materials, and conductive polymers, which can be manufactured with AM processes such as SLS, MJF, and FFF in thermoplastics.

A solution-precipitation approach to prepare the CNTs coated PA12 and TPU powders with the capability of SLS fabrication polymeric composite with an electrically conductive feature was proposed by Yuan et al. [68]. This approach was conducted to provide a flexible CNT/TPU composite to manufacture the electrical strain sensor. Hence, the conductive additives of CNTs mixed with photosensitive resin are employed as feedstock materials to manufacture functional objects with electrical performance [69]. Addition of graphene to the conductive filaments with carbon fiber leads to providing feedstock materials for the FFF process, in which the manufactured objects achieved noticeable electrical conductivity characteristics [70].

For direct ink writing or DIW and droplet printing additive manufacturing process, the groups of conductive inks were conducted. The pattern of the nozzle head can directly form the conductive path of the composite structure. The functionality and performance of manufactured composites are considerably influenced by the strategy of conductive paths. Surfactants functionalize the carbon nanomaterials and the silver/copper nanoparticles and are dispersed into a solvent to provide the ink of DIW and droplet printing additive manufacturing process [71]. These methods and approaches have been conducted for the manufacturing process of different applications such as antenna, strain sensor, and pressure sensor.

The electrically conductive inks can also be manufactured on the flexible surface with the aim of supporting the development of flexible electronics. For instance, instead of conventional layer-wise manufacturing of electrodes, the CNT or graphene blended with polymer additives can be manufactured and 3-D configurated as anode of Li-ion battery [72,73]. The localized distribution and path design of electrically conductive materials in mesoscale can provide exceptional electrical characteristics. For example, controlling the bead size of printing materials and the path design led to monitoring the sensitivity of pressure or strain sensor. It is provided to customize the bandwidth and frequency range in the 3D-printed antenna [74]. The continuous additive manufactured fiber-reinforced composite provides distinctive characteristics to produce a highly conductive path along with fiber orientation. The patterning strategy can provide the 3D conductive path directly to strengthen the electrical conductivity of the composite by alteration of the continuous fiber arrangement within the matrix.

In summary, the electrical conduction of composite can be arranged and fabricated in different scales. The effective method to alter the material formulation and change the microstructure of printed composite in nano/microscale is provided by the addition of nano or micro conductive additives. Also, the advantages in device fabrication and electrical performance in anisotropy are prepared by the conductive path planning and arrangement of continuous fiber in meso/macroscale. The conductivity characteristics of additively manufactured composites are applied in functional device development since most data requires to be transferred to electrical signal to communicate with the external systems.

11.6.2 Electromagnetic Shielding

The electromagnetic metamaterials, which are artificial structures provide astonishing and remarkable interactions with incident waves. These features are not accessible by natural materials. Noticeable identifications of electromagnetic metamaterials are cloak effects, negative refraction index, interference shielding, and tunable wave absorption. These metamaterials are required for various applications including sensing, imaging, non-destructive detection, and electromagnetic shielding.

To realize the functionalities of electromagnetic metamaterials, the key factors are the structural configuration and material composition. An in-depth scientific study is required for the additively manufacturable composite materials/structure to provide desirable characteristics including high absorptivity and wideband wave absorption. The two wave absorbers including the magnetic-dielectric type and electric-dielectric types are composite materials, which consist of magnetic polarized/dielectric phase and conductive/dielectric phase, respectively. The dielectric loss and magnetic loss are influenced by material compositions and their microstructures considerably. The electromagnetic metamaterials wave absorbers have periodic macrostructures, which can monitor the electromagnetic parameters of composites [75].

The fabrication of the continuous CF/PLA composite [76], the graphene composite [77], and honeycomb composites [78] are provided by the FFF additive manufacturing process. The electromagnetic wave absorptivity of 90% in a wideband of 8–18 GHz is provided by the additively manufactured graphene composites with 3D structural grids [77]. The highly effective shielding of CF/PLA is achieved in the range of 6.8–78.9, 51.1–75.6, and 25.1–69.9 dB by adjusting the filling angle (90°–0°), hatch spacing (1.6–0.8 mm), and number of layers (2–12) [78]. The honeycomb composite-based metamaterial absorber was manufactured by combining the 3D-printed plastic dielectric, conductive copper backboard, and resistive patches. The high and wide absorptivity in the band of 4.53–24 GHz are realized and sharp absorption peaks are detected at the frequencies of 19.21, 11.78, and 3.82 GHz by this absorber [78].

Polarization sensitive and wideband metamaterials were manufactured by a hybrid process of FFF and patterning of conductive phases additive manufacturing processes similar to high absorption (across a wide range of frequency) electric excitation conductive plastic surface [79]. The structural composites consisting of PA12 and ferromagnetic materials, which provide a wideband absorption effect in the range of 4–18 GHz, are manufactured by the SLS additive manufacturing process. The SLA additive manufacturing process was applied to provide a gradient refractive-index metamaterial to prepare an electromagnetic wave cloak [80]. An Al_2O_3/SiC_w composite through chemical vapor infiltration on the SLA-manufactured Al_2O_3 structure is developed, which shows an incredible absorption bandwidth ranging from 8.2 to 12.4 GHz [75].

11.7 THERMAL CONDUCTION AND EXPANSION

Thermal conduction and expansion are two important thermal characteristics of composite, which can be altered through designing the microstructural fabrication and the material formulation. The additively manufactured insulative and thermal conductive composites are desirable groups of materials and structures for thermal protection and management in electronics and aerospace [81]. The coefficient of thermal expansion (CTE) of additively manufactured composite structures is adjustable by optimizing and combining the microstructural fabrication that may lead to uncommon thermal expansion characteristics (negative or zero coefficients of expansion) [82,83].

11.7.1 THERMAL CONDUCTIVITY

Additive manufacturing has revolutionized the fabrication of advanced composites and components, offering enhanced design freedom and material customization. However, achieving optimal thermal conductivity remains a crucial challenge, particularly when conventional polymers are employed. The primary contributor to the high thermal resistance observed in such polymers is the dissipation of phonon energy along cross-linked polymer chains.

11.7.1.1 Phonon Propagation and Thermal Conductivity

Phonons play a pivotal role in determining the thermal properties of materials. In additively manufactured composites, phonon propagation within the polymer matrix significantly influences thermal conductivity. As phonons traverse through the composite structure, their energy is dissipated along the complex network of cross-linked polymer chains. Consequently, this energy dissipation leads to noticeable thermal resistance, limiting the overall efficiency of the composite.

11.7.1.2 Enhancing Thermal Conductivity via Fillers

To address the thermal resistance issue, the introduction of thermal conductive fillers is proposed. Graphene-based materials and continuous carbon fibers are identified as promising candidates for enhancing thermal conductivity in additively manufactured composites. These fillers establish an interconnected network throughout the matrix, allowing for improved phonon propagation and heat transfer.

11.7.1.3 Graphene-Based Materials

Graphene, with its exceptional thermal conductivity, represents an attractive choice for enhancing the thermal properties of composites. When incorporated into the polymer matrix, graphene sheets create efficient phonon pathways, resulting in enhanced thermal conductivity. The two-dimensional structure of graphene facilitates superior phonon propagation compared to conventional three-dimensional fillers.

11.7.1.4 Continuous Carbon Fibers

Another compelling option for improving thermal conductivity in composites is the use of continuous carbon fibers. These fibers possess high thermal conductivity along their length, which allows for efficient phonon transport. When properly distributed within the polymer matrix, continuous carbon fibers establish a conductive network that accelerates heat transfer through the composite.

11.7.1.5 Reinforcing Thermal Conduction Characteristics

By integrating graphene-based materials and continuous carbon fibers as thermal conductive fillers, the thermal conduction characteristics of additively manufactured composites are significantly strengthened. The interconnected network formed by these fillers ensures efficient phonon propagation, thereby reducing thermal resistance and enhancing overall thermal conductivity.

In conclusion, this section emphasizes the significance of phonon propagation within the polymer matrix in additively manufactured composites. It underscores the adverse thermal resistance commonly observed in conventional polymers due to phonon energy dissipation. To overcome this limitation, the inclusion of graphene-based materials and continuous carbon fibers as thermal conductive fillers is proposed. The establishment of an interconnected network through these fillers remarkably enhances the thermal conduction characteristics of composites, opening up new possibilities for advanced additive manufacturing applications in various industries.

11.7.2 THERMAL INSULATIVE

Advancements in electronic devices necessitate novel approaches to developing efficient thermal management solutions. In response, researchers have turned their attention towards additively manufactured thermal insulative materials, structures, and composites. While electrically conductive composites have shown relatively high thermal conductivity, the study highlights the potential of highly compressible 3D periodic graphene aerogel microlattices, known for their unique combination of high electrical conduction and a porous structure [81,84].

11.7.2.1 Graphene Aerogel Microlattice

The highly compressible 3D periodic graphene aerogel microlattice is a promising candidate for thermal insulative applications. Its unique structure, composed of interconnected graphene sheets, not only contributes to exceptional thermal insulation but also allows for relatively high electrical conduction. This dual functionality makes it a versatile material for electronic devices where thermal management and electrical conduction are critical aspects.

11.7.2.2 DIW Additive Manufacturing of Silica Aerogel Parts

In this method, a Direct Ink Writing (DIW) additive manufacturing process was utilized to produce miniaturized silica aerogel parts. By printing silica sol, researchers successfully engineered intricate structures with controlled porosity, which is vital for achieving low thermal conductivity. The DIW technique allowed precise control over the material's deposition, resulting in tailor-made thermal insulative components [84].

11.7.2.3 Enhanced Thermal Insulation

The additively manufactured silica aerogel parts exhibited low thermal conductivity in specific regions, validating their potential as highly effective thermal insulators. The presence of porous structures within the aerogel plays a crucial role in hindering heat transfer, while the inherent electrical conduction properties of graphene contribute to overall material performance [84].

Metamaterials with zero or negative CTE hold great promise for applications in various industries, including aerospace, electronics, and engineering. Dual/multiple phases play a critical role in achieving such unique thermal properties, and advances in additive manufacturing techniques have opened new possibilities for their fabrication [83,85].

11.7.2.4 Dual-Material Additive Manufacturing Processes

The metamaterials under investigation can be realized through dual-material selective laser sintering (SLS) or Polyjet additive manufacturing processes. These techniques allow for precise control over the arrangement and distribution of different phases, enabling the creation of complex microstructures with tailored thermal properties.

11.7.2.5 Three-Phase Optimization Topology

The concept of three-phase optimization topology, initially proposed in the mid-1990s, has significantly contributed to the development of materials with high thermal expansion. By strategically combining multiple phases, researchers have been able to design metamaterials that exhibit specific thermal characteristics, including zero or negative CTE [83].

11.7.2.6 Multi-Material Additive Manufacturing Advancements

Recent advancements in multi-material additive manufacturing have revolutionized the digital fabrication of composite structures. This technique allows for the seamless integration of materials with varying CTEs into the microstructural units of metamaterials. By digitally controlling the arrangement of these materials, engineers can tailor the overall thermal properties of the metamaterial to meet specific application requirements.

11.7.2.7 Stretch-Dominated Bi-Material Unit Cells

The work of Xu et al. introduced stretch-dominated bi-material unit cells, which offer a compelling approach to achieving zero/negative CTE in metamaterials. These unit cells combine materials with high and low CTEs, arranged in a diamond-shaped building block structure. The hierarchical lattices formed by these building blocks exhibit exceptional thermal properties [83].

11.7.2.8 Adjustable Structural Performance and CTE

The key to unlocking the full potential of metamaterials lies in carefully designing the unit cell, selecting appropriate material compositions, and assembling the building blocks. These factors directly influence the structural performance and CTE of the metamaterial. By optimizing these parameters, engineers can tailor the thermal behavior of the metamaterial to meet specific application requirements with precision [83].

The development of metamaterials with zero/negative CTE is a topic of great interest due to its potential impact on various industries. Dual/multiple phases in metamaterials, fabricated through multi-material additive manufacturing processes, offer a promising avenue for achieving this objective. The concept of three-phase optimization topology, along with recent advancements in additive manufacturing, has facilitated the digital fabrication of composite structures with unique thermal properties. Stretch-dominated bi-material unit cells, assembled into hierarchical lattices, further enhance the ability to fine-tune the structural performance and CTE of metamaterials. As research continues in this area, the application of metamaterials with tailored thermal behavior is expected to drive innovation in diverse fields.

REFERENCES

[1] V. Shanmugam, D. J. J. Rajendran, K. Babu, S. Rajendran, A. Veerasimman, U. Marimuthu, S. Singh, O. Das, R. E. Neisiany, M. S. Hedenqvist, F. Berto, and S. Ramakrishna, "The mechanical testing and performance analysis of polymer-fibre composites prepared through the additive manufacturing," *Polym. Test.*, 93, 106925 (2021).

[2] J. Chacón, M. Caminero, P. Núñez, E. García-Plaza, I. García-Moreno, and J. Reverte, "Additive manufacturing of continuous fibre reinforced thermoplastic composites using fused deposition modelling: Effect of process parameters on mechanical properties," *Compos. Sci. Tech.*, 181, 107688 (2019).

[3] A. D. Pertuz, S. Díaz-Cardona, and O. A. González-Estrada, "Static and fatigue behaviour of continuous fibre reinforced thermoplastic composites manufactured by fused deposition modelling technique," *Int. J. Fat.*, 130, 105275 (2020).

[4] J. Shang, X. Tian, M. Luo, W. Zhu, D. Li, Y. Qin, and Z. Shan, "Controllable inter-line bonding performance and fracture patterns of continuous fiber reinforced composites by sinusoidal-path 3D printing," *Compos. Sci. Technol.*, 192, 108096 (2020).

[5] Z. Xu, C. S. Ha, R. Kadam, J. Lindahl, S. Kim, H. F. Wu, V. Kunc, and X. Zheng (Rayne), "Additive manufacturing of two-phase lightweight, stiff and high damping carbon fiber reinforced polymer microlattices," *Addit. Manuf.*, 32, 101106 (2020).

[6] R. D. Kadam, "Design and additive manufacturing of carbon-fiber reinforced polymer microlattice with high stiffness and high damping," Master of Science, Virginia Tech, 2019.

[7] X. Zheng, H. Lee, T. H. Weisgraber, M. Shusteff, J. DeOtte, E. B. Duoss, J. D. Kuntz, M. M. Biener, Q. Ge, J. A. Jackson, S. O. Kucheyev, N. X. Fang, and C. M. Spadaccini, "Ultralight, ultrastiff mechanical metamaterials," *Science*, 344(6190), 1373–1377 (2014).

[8] P. Parandoush, L. Tucker, C. Zhou, and D. Lin, "Laser assisted additive manufacturing of continuous fiber reinforced thermoplastic composites," *Mater. Des.*, 131, 186–195 (2017).

[9] R. Matsuzaki, M. Ueda, M. Namiki, T.-K. Jeong, H. Asahara, K. Horiguchi, T. Nakamura, A. Todoroki, and Y. Hirano, "Three-dimensional printing of continuous-fiber composites by in-nozzle impregnation," *Sci. Rep.*, 6(1), 1–7 (2016).

[10] L. Suárez, and M. del Mar Espinosa, "Assessment on the use of additive manufacturing technologies for acoustic applications," *Int. J. Adv. Manuf. Technol.*, 109(9), 2691–2705 (2020).

[11] D. C. Akiwate, M. D. Date, B. Venkatesham, and S. Suryakumar, "Acoustic properties of additive manufactured narrow tube periodic structures," *Appl. Acoust.*, 136, 123–131 (2018).

[12] T. Dupont, P. Leclaire, R. Panneton, and O. Umnova, "A microstructure material design for low frequency sound absorption," *Appl. Acoust.*, 136, 86–93 (2018).

[13] Y. Xie, T.-H. Tsai, A. Konneker, B.-I. Popa, D. J. Brady, and S. A. Cummer, "Single-sensor multispeaker listening with acoustic metamaterials," *Proc. Nat. Acad. Sci.*, 112(34), 10595–10598 (2015).

[14] O. Godbold, R. Soar, and R. Buswell, "Implications of solid freeform fabrication on acoustic absorbers," *Rapid Prototyping Journal*, Conference: Internoise 2016, Hamburg (2016).

[15] F. Setaki, M. Tenpierik, M. Turrin, and A. van Timmeren, "Acoustic absorbers by additive manufacturing," *Build. Environ.*, 72, 188–200 (2014).

[16] U. Berardi, "Destructive interferences created using additive manufacturing," *Can. Acoust.*, 45(3), 44–45 (2017).

[17] A. Arjunan, "Acoustic absorption of passive destructive interference cavities," *Mater. Today Commun.*, 19, 68–75 (2019).

[18] A. Arjunan, "Targeted sound attenuation capacity of 3D printed noise cancelling waveguides," *Appl. Acoust.*, 151, 30–44 (2019).

[19] X. Cai, Q. Guo, G. Hu, and J. Yang, "Ultrathin low-frequency sound absorbing panels based on coplanar spiral tubes or coplanar Helmholtz resonators," *Appl. Phys. Lett.*, 105(12), 121901 (2014).

[20] M. Yang, S. Chen, C. Fu, and P. Sheng, "Optimal sound-absorbing structures," *Mater. Horiz.*, 4(4), 673–680 (2017).

[21] N. Gao, and H. Hou, "Sound absorption characteristic of micro-helix metamaterial by 3D printing," *Theor. Appl. Mech. Lett.*, 8(2), 63–67 (2018).

[22] Z. Liu, J. Zhan, M. Fard, and J. L. Davy, "Acoustic properties of a porous polycarbonate material produced by additive manufacturing," *Mater. Lett.*, 181, 296–299 (2016).

[23] C. Jiang, D. Moreau, and D. Doolan, Acoustic absorption of porous materials produced by additive manufacturing with varying geometrics, 2017.

[24] Z. Liu, J. Zhan, M. Fard, and J. L. Davy, "Acoustic properties of multilayer sound absorbers with a 3D printed micro-perforated panel," *Appl. Acoust.*, 121, 25–32 (2017).

[25] Z. Liu, J. Zhan, M. Fard, and J. L. Davy, "Acoustic measurement of a 3D printed micro-perforated panel combined with a porous material," *Measurement*, 104, 233–236 (2017).

[26] T. G. Zieliński, F. Chevillotte, and E. Deckers, "Sound absorption of plates with micro-slits backed with air cavities: Analytical estimations, numerical calculations and experimental validations," *Appl. Acoust.*, 146, 261–279 (2019).

[27] S. Deshmukh, H. Ronge, and S. Ramamoorthy, "Design of periodic foam structures for acoustic applications: Concept, parametric study and experimental validation," *Mater. Des.*, 175, 107830 (2019).

[28] F. Shen, S. Yuan, Y. Guo, B. Zhao, J. Bai, M. Qwamizadeh, C. K. Chua, J. Wei, and K. Zhou, "Energy absorption of thermoplastic polyurethane lattice structures via 3D printing: Modeling and prediction," *Int. J. Appl. Mech.*, 8(07), 1640006 (2016).

[29] S. Yuan, F. Shen, J. Bai, C. K. Chua, J. Wei, and K. Zhou, "3D soft auxetic lattice structures fabricated by selective laser sintering: TPU powder evaluation and process optimization," *Mater. Des.*, 120, 317–327 (2017).

[30] D. W. Abueidda, M. Bakir, R. K. A. Al-Rub, J. S. Bergström, N. A. Sobh, and I. Jasiuk, "Mechanical properties of 3D printed polymeric cellular materials with triply periodic minimal surface architectures," *Mater. Des.*, 122, 255–267 (2017).

[31] D. Li, D. I. Levin, W. Matusik, and C. Zheng, "Acoustic voxels: Computational optimization of modular acoustic filters," *ACM Trans. Graph. (TOG)*, 35(4), 1–12 (2016).

[32] C. He, S.-Y. Yu, H. Ge, H. Wang, Y. Tian, H. Zhang, X.-C. Sun, Y. B. Chen, J. Zhou, M.-H. Lu, and Y.-F. Chen, "Three-dimensional topological acoustic crystals with pseudospin-valley coupled saddle surface states," *Nat. Commun.*, 9(1), 1–7 (2018).

[33] D. Lavazec, G. Cumunel, D. Duhamel, and C. Soize, "Experimental evaluation and model of a nonlinear absorber for vibration attenuation," *Commun. Nonlinear Sci. Numer. Simul.*, 69, 386–397 (2019).

[34] W. Huang, H. Zhang, D. J. Inman, J. Qiu, C. E. Cesnik, and H. Ji, "Low reflection effect by 3D printed functionally graded acoustic black holes," *J. Sound Vib.*, 450, 96–108 (2019).

[35] B. M. P. Chong, L. B. Tan, K. M. Lim, and H. P. Lee, "A review on acoustic black-holes (ABH) and the experimental and numerical study of ABH-featured 3D printed beams," *Int. J. Appl. Mech.*, 9(06), 1750078 (2017).

[36] S. Algermissen, and H. P. Monner, "Reduction of low-frequency sound transmission using an array of 3D-printed resonant structures," in *Smart Materials, Adaptive Structures and Intelligent Systems*, ASME 2018 Conference on Smart Materials, Adaptive Structures and Intelligent Systems, Vol. 51944, p. V001T04A015, 2018. https://doi.org/10.1115/SMASIS2018-7985

[37] C. Casarini, J. F. Windmill, and J. C. Jackson, "3D printed small-scale acoustic metamaterials based on Helmholtz resonators with tuned overtones," in *2017 IEEE Sensors*, IEEE, Glasgow, pp. 1–3, 2017.

[38] P. Wu, Q. Mu, X. Wu, L. Wang, X. Li, Y. Zhou, S. Wang, Y. Huang, and W. Wen, "Acoustic absorbers at low frequency based on split-tube metamaterials," *Phys. Lett. A*, 383(20), 2361–2366 (2019).

[39] L. Friedrich, and M. Begley, "Printing direction dependent microstructures in direct ink writing," *Addit. Manuf.*, 34, 101192 (2020).

[40] S. Yuan, S. Li, J. Zhu, and Y. Tang, "Additive manufacturing of polymeric composites from material processing to structural design," *Compos. Part B Eng.*, 219, 108903 (2021).

[41] X. Zhang, Z. Qu, X. He, and D. Lu, "Experimental study on the sound absorption characteristics of continuously graded phononic crystals," *AIP Adv.*, 6(10), 105205 (2016).

[42] A. H. Vu, Y.-I. Hwang, H.-J. Kim, and S.-J. Song, "Focusing ultrasonic wave using metamaterial-based device with cross shape," *J. Korean Phys. Soc.*, 74(3), 274–279 (2019).

[43] Y. Xie, Y. Fu, Z. Jia, J. Li, C. Shen, Y. Xu, H. Chen, and S. A. Cummer, "Acoustic imaging with metamaterial Luneburg lenses," *Sci. Rep.*, 8(1), 1–6 (2018).

[44] X. Jia, Y. Li, Y. Zhou, M. Hong, and M. Yan, "Wide bandwidth acoustic transmission via coiled-up metamaterial with impedance matching layers," *Sci. China Phys. Mech. Astron.*, 62(6), 1–8 (2019).

[45] G. Memoli, M. Caleap, M. Asakawa, D. R. Sahoo, B. W. Drinkwater, and S. Subramanian, "Metamaterial bricks and quantization of meta-surfaces," *Nat. Commun.*, 8(1), 1–8 (2017).

[46] Y. Xie, B.-I. Popa, L. Zigoneanu, and S. A. Cummer, "Measurement of a broadband negative index with space-coiling acoustic metamaterials," *Phys. Rev. Lett.*, 110(17), 175501 (2013).

[47] Y. Xie, A. Konneker, B.-I. Popa, and S. A. Cummer, "Tapered labyrinthine acoustic metamaterials for broadband impedance matching," *Appl. Phys. Lett.*, 103(20), 201906 (2013).

[48] Y. Xie, W. Wang, H. Chen, A. Konneker, B.-I. Popa, and S. A. Cummer, "Wavefront modulation and subwavelength diffractive acoustics with an acoustic metasurface," *Nat. Commun.*, 5(1), 1–5 (2014).

[49] W. Wang, Y. Xie, B.-I. Popa, and S. A. Cummer, "Subwavelength diffractive acoustics and wavefront manipulation with a reflective acoustic metasurface," *J. Appl. Phys.*, 120(19), 195103 (2016).

[50] W. Wang, Y. Xie, A. Konneker, B.-I. Popa, and S. A. Cummer, "Design and demonstration of broadband thin planar diffractive acoustic lenses," *Appl. Phys. Lett.*, 105(10), 101904 (2014).

[51] Y. Li, X. Jiang, R.-Q. Li, B. Liang, X.-Y. Zou, L.-L. Yin, and J.-C. Cheng, "Experimental realization of full control of reflected waves with subwavelength acoustic metasurfaces," *Phys. Rev. Appl.*, 2(6), 064002 (2014).

[52] Y. Li, X. Jiang, B. Liang, J.-C. Cheng, and L. Zhang, "Metascreen-based acoustic passive phased array," *Phys. Rev. Appl.*, 4(2), 024003 (2015).

[53] X. Zhu, K. Li, P. Zhang, J. Zhu, J. Zhang, C. Tian, and S. Liu, "Implementation of dispersion-free slow acoustic wave propagation and phase engineering with helical-structured metamaterials," *Nat. Commun.*, 7(1), 1–7 (2016).

[54] S. Liang, T. Liu, F. Chen, and J. Zhu, "Theoretical and experimental study of gradient-helicoid metamaterial," *J. Sound Vib.*, 442, 482–496 (2019).

[55] X. Jiang, Y. Li, B. Liang, J.-C. Cheng, and L. Zhang, "Convert acoustic resonances to orbital angular momentum," *Phys. Rev. Lett.*, 117(3), 034301 (2016).

[56] M. Dah-You, "Theory and design of microperforated panel sound-absorbing constructions," *Sci. Sin.*, 18(1), 55–71 (1975).

[57] X.-L. Gai, T. Xing, X.-H. Li, B. Zhang, Z.-N. Cai, and F. Wang, "Sound absorption properties of microperforated panel with membrane cell and mass blocks composite structure," *Appl. Acoust.*, 137, 98–107 (2018).

[58] K. Sakagami, T. Nakamori, M. Morimoto, and M. Yairi, "Double-leaf microperforated panel space absorbers: A revised theory and detailed analysis," *Appl. Acoust.*, 70(5), 703–709 (2009).

[59] V. Sekar, S. Y. E. Noum, A. Putra, S. Sivanesan, and D. D. C. V. Sheng, "Fabrication of light-weighted acoustic absorbers made of natural fiber composites via additive manufacturing," *Int. J. Lightweight Mater. Manuf.*, 5(4), 520–527 (2022).

[60] S. Yuan, F. Shen, C. K. Chua, and K. Zhou, "Polymeric composites for powder-based additive manufacturing: Materials and applications," *Progr. Polym. Sci.*, 91, 141–168 (2019).

[61] B. Ravishankar, S. K. Nayak, and M. A. Kader, "Hybrid composites for automotive applications-A review," *J. Reinf. Plast. Compos.*, 38(18), 835–845 (2019).

[62] M. A. Fentahun, and M. Savas, "Materials used in automotive manufacture and material selection using ashby charts," *Int. J. Mater. Eng.*, 8(3), 40–54 (2018).

[63] O. Akampumuza, P. M. Wambua, A. Ahmed, W. Li, and X. H. Qin, "Review of the applications of biocomposites in the automotive industry," *Polym. Compos.*, 38(11), 2553–2569 (2017).

[64] I. Kaur, and P. Singh, "State-of-the-art in heat exchanger additive manufacturing," *Int. J. Heat Mass Trans.*, 178, 121600 (2021).

[65] D. C. Deisenroth, R. Moradi, A. H. Shooshtari, F. Singer, A. Bar-Cohen, and M. Ohadi, "Review of heat exchangers enabled by polymer and polymer composite additive manufacturing," *Heat Trans. Eng.*, 39(19), 1648–1664 (2018).

[66] U. Fasel, D. Keidel, L. Baumann, G. Cavolina, M. Eichenhofer, and P. Ermanni, "Composite additive manufacturing of morphing aerospace structures," *Manuf. Lett.*, 23, 85–88 (2020).

[67] Q. Zhao, K. Zhang, S. Zhu, H. Xu, D. Cao, L. Zhao, R. Zhang, and W. Yin, "Review on the electrical resistance/conductivity of carbon fiber reinforced polymer," *Appl. Sci.*, 9(11), 2390 (2019).

[68] S. Yuan, Y. Zheng, C. K. Chua, Q. Yan, and K. Zhou, "Electrical and thermal conductivities of MWCNT/polymer composites fabricated by selective laser sintering," *Compos. Part A Appl. Sci. Manuf.*, 105, 203–213 (2018).

[69] G. Gonzalez, A. Chiappone, I. Roppolo, E. Fantino, V. Bertana, F. Perrucci, L. Scaltrito, F. Pirri, and M. Sangermano, "Development of 3D printable formulations containing CNT with enhanced electrical properties," *Polymer*, 109, 246–253 (2017).

[70] A. Dorigato, V. Moretti, S. Dul, S. Unterberger, and A. Pegoretti, "Electrically conductive nanocomposites for fused deposition modelling," *Synth. Met.*, 226, 7–14 (2017).

[71] J. H. Kim, S. Lee, M. Wajahat, H. Jeong, W. S. Chang, H. J. Jeong, J.-R. Yang, J. T. Kim, and S. K. Seol, "Three-dimensional printing of highly conductive carbon nanotube microarchitectures with fluid ink," *ACS Nano*, 10(9), 8879–8887 (2016).

[72] X. Tian, J. Jin, S. Yuan, C. K. Chua, S. B. Tor, and K. Zhou, "Emerging 3D-printed electrochemical energy storage devices: A critical review," *Adv. Energy Mater.*, 7(17), 1700127 (2017).

[73] B. Chang, X. Li, P. Parandoush, S. Ruan, C. Shen, and D. Lin, "Additive manufacturing of continuous carbon fiber reinforced poly-ether-ether-ketone with ultrahigh mechanical properties," *Polym. Test.*, 88, 106563 (2020).

[74] D. Zhang, B. Chi, B. Li, Z. Gao, Y. Du, J. Guo, and J. Wei, "Fabrication of highly conductive graphene flexible circuits by 3D printing," *Synth. Met.*, 217, 79–86 (2016).

[75] H. Mei, X. Zhao, S. Zhou, D. Han, S. Xiao, and L. Cheng, "3D-printed oblique honeycomb Al2O3/SiCw structure for electromagnetic wave absorption," *Chem. Eng. J.*, 372, 940–945 (2019).

[76] L. Yin, X. Tian, Z. Shang, X. Wang, and Z. Hou, "Characterizations of continuous carbon fiber-reinforced composites for electromagnetic interference shielding fabricated by 3D printing," *Appl. Phys. A*, 125, 1–11 (2019).

[77] L. Yin, X. Tian, Z. Shang, and D. Li, "Ultra-broadband metamaterial absorber with graphene composites fabricated by 3D printing," *Mater. Lett.*, 239, 132–135 (2019).

[78] W. Jiang, L. Yan, H. Ma, Y. Fan, J. Wang, M. Feng, and S. Qu, "Electromagnetic wave absorption and compressive behavior of a three-dimensional metamaterial absorber based on 3D printed honeycomb," *Sci. Rep.*, 8(1), 1–7 (2018).

[79] Y. Lu, B. Chi, D. Liu, S. Gao, P. Gao, Y. Huang, J. Yang, Z. Yin, and G. Deng, "Wideband metamaterial absorbers based on conductive plastic with additive manufacturing technology," *ACS Omega*, 3(9), 11144–11150 (2018).

[80] L. Yin, J. Doyhamboure, X. Tian, and D. Li, "Design and characterization of radar absorbing structure based on gradient-refractive-index metamaterials," *Compos. Part B Eng.*, 132, 178–187 (2018).

[81] S. Zhao, G. Siqueira, S. Drdova, D. Norris, C. Ubert, A. Bonnin, S. Galmarini, M. Ganobjak, Z. Pan, S. Brunner, G. Nyström, J. Wang, M. M. Koebel, and W. J. Malfait, "Additive manufacturing of silica aerogels," *Nature*, 584(7821), 387–392 (2020).

[82] J. B. Hopkins, K. J. Lange, and C. M. Spadaccini, "Designing microstructural architectures with thermally actuated properties using freedom, actuation, and constraint topologies," *J. Mech. Des.*, 135(6), 061004 (2013).

[83] H. Xu, A. Farag, and D. Pasini, "Multilevel hierarchy in bi-material lattices with high specific stiffness and unbounded thermal expansion," *Acta Mater.*, 134, 155–166 (2017).

[84] C. Zhu, T. Yong-Jin Han, E. B. Duoss, A. M. Golobic, J. D. Kuntz, C. M. Spadaccini, and M. A. Worsley, "Highly compressible 3D periodic graphene aerogel microlattices," *Nat. Commun.*, 6(1), 6962 (2015).

[85] Q. Wang, J. A. Jackson, Q. Ge, J. B. Hopkins, C. M. Spadaccini, and N. X. Fang, "Lightweight mechanical metamaterials with tunable negative thermal expansion," *Phys. Rev. Lett.*, 117(17), 175901 (2016).

12 Additive Manufacturing in Composite

Applications and Models

12.1 INTRODUCTION

Polymer-based composites provide a considerable type of materials with the potential for employment in AM technology. Various forms of polymers including liquid and thermoplastic melts have been considered in the AM technology. In addition, the particle reinforcements are also considered with the aim of improving the polymer's properties such as PLA (polylactic acid) [1,2], ABS (butadiene styrene) [3,4], and PC (polycarbonate) [5] along with thermosetting resins such as epoxy polymers. Studies revealed that ABS and PLA are two key polymers in AM composites [6]. Some efforts were performed on the development of composites with short reinforcement [7]. Adding carbon nanotubes (CNTs) as an additive in ABS polymer-based material and manufacturing with FDM process leads to an increase of 31% in tensile strength based on study [8]. Other additive materials such as glass, iron, and copper particles were considered in AM process with the aim of improvement of the tensile modulus [9–11]. Adding particles such as ceramic [12], tungsten [13], and aluminum oxide [11] to polymer resin in the AM process provides improvement of dielectric permittivity and wear resistance.

Considering the importance of biomedical applications for human beings, some studies have been focused on additively manufactured scaffolds with a focus on bone tissues [14]. It is reported that there is a lack of study in the field of tissue engineering with respect to other tissues like cardiac tissue. The study in this field is essential because different kinds of tissue replacements need different specifications such as specific pore size, specific mechanical characteristics, and specific morphologies. The additively manufactured polymers reinforced with fibers and particles were applied in different applications. The summary of various techniques, particle-reinforced polymer composites manufactured with additive manufacturing processes including FDM and SLA, and improved characteristics are provided in Table 12.1. Some of the manufactured composites with additive manufacturing techniques FDM and some of the applications are reported in Table 12.2.

12.2 AM MECHANICAL APPLICATIONS (2021-VP)

Stereolithography or SLA additive manufacturing process has been considered to process photosensitive materials for different applications such as mechanical. The method of selection of processing parameters and post-curing treatments are

          DOI: 10.1201/9781003429197-12

TABLE 12.1

The Summary of Additive Manufacturing Techniques with Improved Physical Characteristics

Technique	Additively Manufactured Composites	Improved Characteristics	References
FDM	TiO$_2$/ABS	Tensile strength	[15]
	Carbon nanofibers/ABS	Physical characteristics	[8]
	Montmorillonite/ABS	Tensile strength, modulus, flexural strength and modulus, thermal stability	[16]
	Graphene/ABS	Electrical and thermal conductivity	[17]
	Carbon nanofiber/polystyrene/ graphite	Electrical characteristics	[18]
SLA	Epoxy/CNT	Reduction of elongation and improve of tensile strength	[19]
	Graphene oxide/photopolymer	Tensile strength and modulus	[20]
	TiO$_2$/epoxy acrylate	Hardness, flexural strength, tensile strength and modulus	[21]
	BaTiO$_3$/PEGDA	Piezoelectric	[22]
	CNT/acrylic ester	Electromagnetic energy absorption coefficient	[23]
	BST/epoxy	Thermal conductivity	[24]
	Carbon black/nylon-12	Electrical conductivity	[25]
	TiO$_2$/nylon-12 and graphite/ nylon-12	Tensile modulus and reduction of elongation	[26]
	Silica/nylon-11	Increase of tensile and compressive characteristics	[27]

TABLE 12.2

Manufactured Composites with Additive Manufacturing Techniques and Some Applications

Technique	Additively Manufactured Composites	Applications	References
FDM	HA/PLA	Biodegradable scaffold with improved crack resistance during cyclic loading	[28]
	HA/TCP	Scaffold with improved compressive strength	[29]

essential in the process, and they directly influence on the final produced parts. Several studies focused on the effect of processing parameters on manufactured quality and mechanical characteristics of produced objects, which will be discussed in this section.

The impact of layer thickness, build orientation, size, and strain rate effect on the mechanical characteristics of manufactured products with SLA additive manufacturing process was investigated by Naik and Kiran [30]. The improvement of average tensile strength and Young's modulus by 45.17% and 47.38%, respectively, with the rise in strain rate from 0.0033 to 0.0131 per second was indicated. Additively manufactured specimens with a "flat" configuration provide lower Young's modulus and ultimate strength and increase in layer thickness from 25 to 100 μm lead to an increase of 30.15% and 21.22% Young's modulus and tensile strength, respectively.

Some of the previous studies on different polymer-based composites, preceramic polymers, and polymers manufactured by the SLA additive manufacturing process for mechanical applications are provided in Table 12.3 based on base material, additives and reinforcement, and improved characteristics.

As can be seen in Table 12.3, different base materials including methacrylate-based photopolymer resin, bi- and trifunctional acrylates, photopolymerizable resin, formlabs standard clear resin, preceramic polymers, liquid methylvinyl-hydrogen polysiloxane, allylhydopolycarbosilane, epoxy-based photocurable resin, acrylate photoactive resin, monomers such as iso-bornyl acrylate and bi-sphenol-A-ethoxylate diacrylate are considered in SLA process for mechanical applications. The applied additives and reinformed, which have been applied in SLA process, are photoinitiator, pigments, Al_2O_3, 3DMix ZrO_2, Methacrylic acid esters, photo absorber, free radical scavenger: Hydroquinone, Nil, light absorber, and surface reduction aging. Improved mechanical characteristics in SLA additive manufacturing were an increase of elastic modulus, Young's modulus, compressive strength, impact strength, and a decrease in coefficient of variance.

Digital light processing or DLP is fairly like stereolithography or SLA, the difference is that SLA applied laser beam technology to cure the layers by point-to-point tracing the geometry and provide better manufactured quality and accuracy. Demands for design and manufacturing industries for manufacturing products with DLP are growing. It is because the fast-manufacturing speed and high resolution of DLP additive manufacturing process and like SLA process, processing parameters play a critical role on the mechanical characteristics such as UTS, ultimate elastic modulus, and ultimate strain of fabricated products.

Different studies have been focused on polymer-based composites manufactured by the DLP additive manufacturing process with consideration of mechanical applications. The effect of processing parameters such as part orientation, exposure time, and layer thickness on mechanical characteristics was considered and noticeable results were indicated [37]. It is revealed that a decrease of layer thickness from 50 to 10 μm and an increase in exposure time from 1.6 to 2 seconds provide an increase in mechanical characteristics. Some of the related studies to different polymer-based composites manufactured from DLP additive manufacturing process for mechanical applications are provided in Table 12.4, which considers the base materials, additives and reinforcement, and improved mechanical characteristics.

As reported in Table 12.4, various base materials were considered in DLP additive manufacturing process including PR-48, Autodesk Standard clear resin, SUV elastomer polymer resin, Calcium phosphate, commercial photocurable

TABLE 12.3

Different Polymer Based Composites and Polymers Produced via the SLA Process for Mechanical Applications

Base Material	Additives and Reinforcement	Improved Characteristics	References
Methacrylate-based photopolymer resin	• Photo initiator • Pigments	• Improving tensile strength • Improving Young's modulus • Improving strain rate	[30]
Bi- and trifunctional acrylates	• Al_2O_3	• Maximum dimensional accuracy of 45%	[31]
Photopolymerizable resin	• 3DMix ZrO_2	• Fracture resistance through nonaging process • Maximum flexural strength	[32]
Formlabs standard clear resin	• Methacrylic acid esters • Photo initiator	• Maximum elastic modulus of 1.2 GPa • Maximum UTS of 31 MPa • Maximum ultimate strain 34%	[33]
Preceramic polymers			
• Liquid methylvinylhydrogen polysiloxane	• Photo initiator: Phenylbis phosphine oxide	• Hardness of 12 GPa • Mean reduced modulus of 106 GPa	[34]
• Allylhydopolycarbosilane	• Photo absorber: Sudan Orange G • Free radical scavenger: Hydroquinone	• Compressive strength of 216 MPa	[35]
• Epoxy based photocurable resin	• Nil	• Maximum Young's modulus of 908 MPa • Maximum impact strength of 22.8 kJ/m^2 • Maximum compressive strength of 25.9 MPa	[35]
• Acrylate photoactive resin	• Nil	• Decrease in percent deviation • Decrease in coefficient of variance • Elastic modulus ratios range of 0.81–0.97	[36]
Monomers			
• Iso-bornyl acrylate • Bi-sphenol-A-ethoxylate diacrylate	• Photo initiator • Light absorber • Surface reduction aging	• Maximum elastic modulus 540 MPa	[36]

TABLE 12.4

Different Polymers Produced via the DLP Process for Mechanical Applications

Base Material	Additives and Reinforcement	Improved Characteristics	References
PR-48, Autodesk Standard clear resin	• Nil	• Increase of elastic modulus 214% • Increase of ultimate tensile strength 301% • Increase ultimate strain 347%	[37]
		• Maximum elastic modulus 0.75 GPa • Maximum ultimate stress 21.7 MPa • Maximum ultimate strain 5.8	[38]
SUV elastomer polymer resin	• Photo initiator • Monomer • Crosslinker • Isobornyl acrylate	• Increase of Young's modulus 625% • Increase of elongation 358% • Increase of toughness	[39]
Calcium phosphate	• Camphor	• Flexural strength 87 MPa • Reduction of viscosity • Decrease in shrinkage • Increase in compressive strength	[40]
Commercial photocurable acrylic-based resins	• Acrylic monomers • Oligomers • Photo initiator • Inorganic additives	• Maximum tensile strength 53 MPa • Maximum compressive strength 110 MPa • Maximum flexural strength 79 MPa	[41]
Ebecryl 8413	• Monomers • Photo initiator	• Tensile strength 10 MPa • Maximum elastic modulus 10 MPa • Maximum elongation 365%	[42]
Photoresin G+Yellow	• Nil	• Elastic modulus 64 MPa • Maximum offset compressive strength 1.98 MPa • Maximum strain-to-break 0.097 mm/mm	[43]

(Continued)

TABLE 12.4 (Continued)
Different Polymers Produced via the DLP Process for Mechanical Applications

Base Material	Additives and Reinforcement	Improved Characteristics	References
Bis(propylacrylamide)poly	• Photo initiator • Toluene	• Maximum elongation 472%	[44]
VisiJet FTX green	• Monomers • Triethylenglycol diacrylate • Tricyclodecane dimethanol diacrylate • Photo initiator • Phenyl oxide bis phosphine	• Increase of tensile modulus 40% • Increase of ultimate flexural strength 77% • Maximum elastic modulus 1,073 MPa	[45]
Epoxy silicone resin	• Catalyst moderator • Additive • Photo initiator • Absorber	• Maximum compressive strength 3.45 MPa	[46]
Methacrylated PDMS-macromer	• Photo initiator	• Maximum elastic modulus 0.96 MPa • Maximum elongation 66.37% • Maximum UTS 0.49 MPa	[47]
Acrylate-based photosensitive resins	• Photo initiator • Monomer • Crosslinker	• Increase in storage modulus 59%	[48]
Acrylic based photocurable resin	• Multi-walled carbon nanotubes	• Increase in UTS 21% • Increase in Young's modulus 45% • Decrease of elongation 10%	[49]
3DM-ABS, VeroClear, VeroWhite, TangoPlus and Veroblack	• Nil	• Maximum Young's modulus 129.9 Mpa	[50]

(Continued)

TABLE 12.4 (Continued)
Different Polymers Produced via the DLP Process for Mechanical Applications

Base Material	Additives and Reinforcement	Improved Characteristics	References
Urethane acrylate resin Ebecryl 8232	• Fe_3O_4 nanofillers	• Glass transition temperature T_g 10.7°C • Maximum elastic modulus 7 MPa	[51]
Tangoplus FLX 930	• Photo absorber	• Maximum elongation 170%–220%. • Increase of ultimate strain 246%	[52]
Polysilazane	• Binder • Photo initiator • Initiator	• Maximum compressive strength 65.5 MPa • Maximum elastic modulus 768.5 MPa • Linear shrinkage 27.63% • Relative density 88.43%	[53]
Bisphenol A ethoxylate diacrylate	• Photo initiator • Photo absorber • Poly Glycidyl methacrylate • N-buttyl acrylate	• Increase of Young's modulus to 1.2 MPa • Glass transition temperature T_g range from 14°C to 68°C	[54]
CLRS, photocurable resins	• Zirconium • Isopropanol • High density polyethylene	• Maximum compressive strength 428.1 MPa • Maximum flexural strength 129.5 MPa	[55]
Highly acrylated, liquid photocurable polysiloxane	• Dispersing aging • Photo initiator • Photo absorber	• Maximum compressive strength of 1.8 MPa • Total porosity 90 vol%	[56]
Commercial photocurable acrylic resin	• Lunar regolith CLRS-2 solid loading	• Maximum compressive strength 312.2 MPa • Maximum flexural strength 74.1 MPa • Maximum hardness 822 HV	[57]

(Continued)

TABLE 12.4 (Continued)
Different Polymers Produced via the DLP Process for Mechanical Applications

Base Material	Additives and Reinforcement	Improved Characteristics	References
Acrylate-polydimethysiloxane	• Oligomer • Photo initiator • Dyeing agent	• Maximum UTS 1.5 MPa • Maximum Young's modulus 0.70 MPa	[58]
Sylgard 184 polydimethylsiloxane pre-polymer Monoclinic ZeO_2 with solid loading of 50 vol%	• S1813 particles • 3Y-TZP • Monomers • Photo initiator • Dispersant • Sintering additive	• Maximum Young's modulus 8.19 MPa • Maximum flexural strength 306.5 MPa • Linear shrinkage < 20% • Relative density 96.40%	[59] [60]
Ceramic powders	• Photo initiator • Ethoxylated pentaerythritol tetraacrylate • 1,6-Hexanediol diacrylate • Di-functional aliphatic urethane acrylate • Octanol • Poly-ethylene glycol	• Maximum flexural strength 398 MPa • Sintered density 99%	[61]
Multifunctional acrylate monomers	• Partially stabilized zirconia • Photo initiator • Dispersant	• Maximum flexural strength 803 MPa • Maximum Young's modulus 203 MPa	[62]
3 mol% yttrium-stabilized zirconia (3Y-TZP) suspension	• Nil	• Maximum flexural strength 580 MPa	[63]

(Continued)

TABLE 12.4 (*Continued*)
Different Polymers Produced via the DLP Process for Mechanical Applications

Base Material	Additives and Reinforcement	Improved Characteristics	References
Polyethylene glycolAcryl@PEG4000	• Diluent • Photo initiator • Polymerization inhibitor Orasol dye Acryl@PEG4000	• Maximum tensile strength 17.2 MPa • Maximum residual strain 19.0% • Maximum elastic modulus 333.4 MPa	[63]
Polyurethane acrylate photopolymer	• Acrylate oligomer	• Maximum tensile strength • Maximum elongation	[64]
Cordierite powder	• Diluent/monomer TMPTA • Diluent/monomer HDDA • Surfactant KH-570 • Dispersant disperbyk-111 • Photo initiator	• Maximum elastic modulus • Maximum compressive strength 46.3 MPa	[64]
Shrinkage calcined kaolin powder solid loading of 45 vol%	• Monomer • Crosslinker • Photo initiator	• Maximum transvers rupture strength 9.98 MPa • Maximum porosity 24.5%	[65]
2-Hydroxyethyl acrylate	• Photo initiator • Light absorbers	• Maximum strain range 156%–188%	[66]
Photosensitive resin	• AlN and Y_2O_3 powder having mass ratio of 95:5 and solid loading of 55 vol%	• Increase of flexural strength 90% • Maximum bending strength 265 MPa • Total porosity 4% • Maximum thermal conductivity 155 W/m·K	[65]
Photosensitive resin	• Zirconia	• Maximum compressive strength 105 MPa	[67]
Methacrylate resins	• Photo initiators	• Maximum yield strength 41 MPa	[68]

acrylic-based resins, Ebecryl 8413, Photoresin G + Yellow, Bis(propylacrylamide) poly, VisiJet FTX green, epoxy silicone resin, Methacrylated PDMS-macromer, Acrylate-based photosensitive resins, Acrylic based photocurable resin, 3DM-ABS, VeroClear, VeroWhite, TangoPlus and Veroblack, Urethane acrylate resin Ebecryl 8232, Tangoplus FLX 930, polysilazane, Bisphenol-A-ethoxylate diacrylate, CLRS, photocurable resins, highly acrylated, liquid photocurable polysiloxane, commercial photocurable acrylic resin, Acrylate-polydimethysiloxane, Sylgard 184 polydimethylsiloxane pre-polymer, Monoclinic ZeO_2 with solid loading of 50 vol%, ceramic powders, multifunctional acrylate monomers, 3 mol% yttrium-stabilized zirconia (3Y-TZP) suspension, polyethylene glycolAcryl@ PEG4000, polyurthane acrylate photopolymer, Cordierite powder, Shrinkage calcined kaolin powder solid loading of 45 vol%, 2-hydroxyethyl acrylate, photosensitive resin, photosensitive resin, and methacrylate resins.

Although various additives and reinforcement can be considered in DLP process for mechanical applications, but studies revealed materials such as Nil, photoinitiator, monomer, crosslinker, isobornyl acrylate, camphor, acrylic, monomers, oligomers, photoinitiator, inorganic additives, toluene, triethylenglycol diacrylate, tricyclodecane dimethanol diacrylate, phenyl oxide bis-phosphine, catalyst moderator, multi-walled carbon nanotubes, Fe_3O_4 nanofillers, photo absorber, binder, glycidyl methacrylate, N-buttyl acrylate, zirconium, isopropanol, high-density polyethylene, dispersing aging, lunar regolith CLRS-2 solid loading, oligomer, dyeing agent, S1813 particles, 3Y-TZP, dispersant, sintering additive, ethoxylated pentaerythritol tetraacrylate, 1,6-hexanediol diacrylate, di-functional aliphatic urethane acrylate, octanol, poly-ethylene glycol, partially stabilized zirconia, polymerization inhibitor, Acryl@PEG4000, acrylate oligomer, diluent/monomer TMPTA, diluent/monomer HDDA, surfactant KH-570, dispersant disperbyk-111, AlN and Y_2O_3 powder having mass ratio of 95:5 and solid loading of 55 vol%, zirconia.

Different aims can be considered to produce parts with additive manufacturing process. Considering Table 12.4, increase in elastic modulus, ultimate tensile strength, ultimate strain, Young's modulus, toughness, elongation, compressive strength, flexural strength, glass transition temperature, and reduction of viscosity, shrinkage were achieved by polymer-based composites produced by DLP process for mechanical applications.

12.3 BIOMEDICAL APPLICATIONS

Additive manufacturing process has been extensively considered for biomedical and dentistry applications. This is because AM processes provide the freedom to design patient-specific and customized dentures, in which processing parameters optimization is essential to provide required product performance. Therefore, mass studies have been focused on additively manufactured product with the aim of biomedical applications.

Manufacturing the denture base with SLA additive manufacturing process through CAD model was reported by Hada et al. [69]. The desired dentures were fabricated in three different directions with the aim of analyzing the effect of print direction on stress distribution. It is reported that the lowest stress distribution is provided for dentures at a direction with an angle of 45°. Since photopolymer resin has a cytotoxic nature, some studies presented synthesizing biocompatible and biodegradable materials to utilize in additive manufacturing process. Applying mask projection micro-stereolithography, biocompatible and biodegradable tissues with low cytotoxicity, improved printability, and proper cell adhesion is studied [70]. The feature size of 50 μm with PTEGA-GMA polyester with mouse preosteoblasts was provided, which has biocompatibility and adhesion characteristics. The summary of additively manufactured polymer-based composites via SLA and DLP processes for biomedical applications is provided in Tables 12.5 and 12.6, respectively.

Considering Tables 12.5 and 12.6, different results can be concluded from SLA and DLP process manufactured biomedical applications. First, methacrylate-based photopolymer resin, poly triethylene glycol adipate dimethacrylate, epoxidized sucrose soyate for SLA process and Yttria tetragonal stabilized zirconia, poly(1,3-propanediyl-co-glyceryl), glycidyl methacrylate, polyerthylene, 1,6-hexanediol diacrylate, bisphenol A ethoxylate diacrylate, PCL-based polyurethane acrylates, m-PUA530, m-PUA800 and m-PUA1000, acrylic resin, acrylated polymer resin (tri-propylene glycol diacrylate) resin, poly-(ethylene glycol) diacrylate, polyethylene glycol diacrylate, DSM's SomosO WaterShed XC, hydroxyapatite (HA), Zirconia powder, Yttria-stabilized, zirconia (Zpex4), CaP powder, poly L-lactic acid (PLLA) having acrylate resin (in form of oligomeric precursors, RF resin-HT), etc. for DLP process are considered as base materials. Second, various materials can be used as additives and reinforcement for SLA (such as diphenyl phosphine oxide, photoinitiator, avobenzone, MC3T3-E1 mouse preosteoblasts, diluent, and monomer) and DLP (like diethyl fumarate (DEF), phenylbis(2,4,6-trimethylbenzoyl), phosphine oxide (BAPO) and Irgacure, oxybenzone (HMB), 20 wt% TMPTMA, 20 wt% N-methyl pyrrolidone, TPO (4 wt%), orange dye,

TABLE 12.5

Different Polymers Produced via the SLA Process for Biomedical Applications

Base Material	Additives and Reinforcement	Improved Characteristics	References
Methacrylate-based photopolymer resin	• Diphenyl phosphine oxide	• The minimum stress distribution at 45°	[69]
Poly triethylene glycol adipate dimethacrylate	• Photo initiator • Avobenzone • MC3T3-E1 mouse preosteoblasts	• Maximum storage modulus 11.3 MPa • Glass transition temperature 3.6°C	[70]
Epoxidized sucrose soyate	• Photo initiator • Diluent • Monomer	• Maximum flexural modulus 1,000 Mpa • Glass transition temperature 69.2°C	[71]

TABLE 12.6
Different Polymers Produced via the DLP Process for Biomedical Applications

Base Material	Additives and Reinforcement	Improved Characteristics	References
Yttria tetragonal stabilized zirconia	• Pigment • Al_2O_3 • SiO_2 • Fe_2O_3 • Na_2O	• Maximum flexural strength 943.26 MPa • Maximum characteristic strength 10,006 MPa • Obtained density 99.9%	[72]
Poly(1,3-propanediyl-co-glyceryl)	• Initiators • Terminator • Dyes	• Maximum tensile strength 5.4 MPa • Maximum tensile modulus 61 MPa • Maximum elongation 18% • Maximum flexural modulus 350 MPa • Maximum flexural strength 12.6 MPa	[73]
Glycidyl methacrylate	• Photo initiator	• Maximum compressive stress 910 kPa • Increase of compressive strain up to 80% • Increase of elongation 124%	[74]
Monomers: • Polyerthylene • 1,6-Hexanediol diacrylate • Bisphenol A ethoxylate diacrylate	• Photoinitiator: phosphine oxide-based compound	• Maximum achieved storage modulus around 10^6 GPa	[75]
PCL-based polyurethane acrylates, m-PUA530, m-PUA800 and m-PUA1000	• Polyethylene glycol diacrylate • Polypropylene glycol • 3% w/w diphenyl phosphine	• Maximum compressive strength of 256.52 MPa • Maximum compressive modulus 34.43 MPa • Maximum compressive strength range 6.8–23.4 MPa	[76]
Acrylic resin	• Commercial powders β-TCP • Pigment	• Maximum achieved compressive strength 22±4 MPa	[77]

(Continued)

TABLE 12.6 (*Continued*)
Different Polymers Produced via the DLP Process for Biomedical Applications

Base Material	Additives and Reinforcement	Improved Characteristics	References
Acrylated polymer resin (tri-propylene glycol diacrylate) resin	• Glass: wollastonite-diopside having ~52 mol% • Wollastonite and 48 mol% diopside	• Total achieved porosity 83 vol% • Maximum achieved mechanical strength 3.2 MPa • Maximum observed linear shrinkage 25%	[78]
Poly-(ethylene glycol) diacrylate	• Photoinitiator: lithium phenyl-2,4,6-trimethylbenzoyiposphinate	• Maximum achieved elastic modulus 3,410 kPa • Maximum achieved ultimate stress 11.5 kPa • Maximum achieved compression strain 25%	[79]
Polyethylene glycol diacrylate	• Photoinitiator: diphenylphosphine oxide • Poly(ethylene glycol) 400 • Sodium chloride (NaCl) • Mannitol	• Maximum achieved tensile strength >1.7 MPa	[80]
DSM's SomosO·WaterShed XC 11122	• Hydroxyapatite powder (10, 20, 30, 40, 45 wt%)	• Maximum achieved compressive strength 12.8 MPa	[81]
Hydroxyapatite (HA)	• Dispersant: liquid sodium polyacrylate	• Maximum achieved compressive strength 15.25 MPa • Maximum achieved compressive modulus 0.97 GPa • Maximum achieved bending strength 41.3 MPa • Total porosity 49.8%	[82]
Zirconia powder, Yttria-stabilized zirconia (Zpex4)	• Monomer: 1,6-hexanediol diacrylate (HDDA) • Diluent: decalin • Dispersant: DISPERBYK180 • Photoinitiator: TPO	• Maximum achieved flexural strength 831 ± 74 MPa • high optical transmittance of 30 (±1.2)%	[83]

(Continued)

TABLE 12.6 (Continued)

Different Polymers Produced via the DLP Process for Biomedical Applications

Base Material	Additives and Reinforcement	Improved Characteristics	References
CaP powder	• Monomer: 1,6-hexanediol diacrylate • Photoinitiator: Phenylbis(2,4,6-trimethylbenzoyl)-phosphine oxide • Surfactants: stearic acid, oleic acid, sebacic acid, and monoalcohol ethoxylate phosphate (MAEP)	• Increase in compressive strength up to 151% • Reduction in porosity up to 12%	[84]
Poly L-lactic acid (PLLA) having 56 wt%	• Cross-linker: 20 wt% TMPTMA • Non-reactive diluent: 20 wt% *N*-methyl pyrrolidone • Photoinitiator: TPO (4 wt%) orange dye	• Maximum achieved compressive strength 2.2 MPa	[85]
Acrylate resin (in form of oligomeric precursors, RF Resin-HT)	• Non-photocurable liquid silicone (H62C) (59 wt% solid loading) • Monomer resins	• Maximum achieved compressive strength 3.7 MPa, • Total porosity 76%	[85]
Hydroxyapatite (HA)	• 1,6-Hexanediol diacrylate • 2-Hydroxyethyl methacrylate (HEMA) • Trimethylolpropane triacrylate (TMPTA) • Photoinitiator: • Diphenyl(2,4,6-trimethyl-benzoyl) phosphine oxide (TPO) • Dispersant: Solsperse 17000	• Maximum achieved compression strength 16.9 MPa • Maximum achieved flexural strength 18.3 MPa • Shrinkage rate 15.3%	[86]
Poly(propylene fumarate) (PPF)	• Diluent: diethyl fumarate (DEF) • Photoinitiators: phenylbis (2,4,6-trimethylbenzoyl) • Phosphine oxide (BAPO) and Irgacure • Radical scavenger: oxybenzone (HMB)	• Maximum achieved elastic modulus 199 MPa at 2.0 kDa • Maximum achieved strain range 15%–25%	[87]

non-photocurable liquid silicone (H62C) (59 wt% solid loading)) processes with the aim of biomedical applications. Third, although different characteristics can be improved with additive manufacturing processes, in biomedical applications, characteristics like Young's modulus, strain range, shrinkage rate, flexural strength, compressive strength, porosity, and tensile strength are investigated in previous studies.

REFERENCES

[1] P. Tran, T. D. Ngo, A. Ghazlan, and D. Hui, "Bimaterial 3D printing and numerical analysis of bio-inspired composite structures under in-plane and transverse loadings," *Compos. Part B Eng.*, 108, 210–223 (2017).

[2] R. Melnikova, A. Ehrmann, and K. Finsterbusch, "3D printing of textile-based structures by fused deposition modelling (FDM) with different polymer materials," *IOP Conf. Ser. Mater. Sci. Eng.*, 62(1), 012018 (2014), IOP Publishing.

[3] B. Tymrak, M. Kreiger, and J. M. Pearce, "Mechanical properties of components fabricated with open-source 3-D printers under realistic environmental conditions," *Mater. Des.*, 58, 242–246 (2014).

[4] Q. Sun, G. Rizvi, C. Bellehumeur, and P. Gu, "Effect of processing conditions on the bonding quality of FDM polymer filaments," *Rapid Prototyp. J.*, 14, 72–80 (2008).

[5] J. Dou, Q. Zhang, M. Ma, and J. Gu, "Fast fabrication of epoxy-functionalized magnetic polymer core-shell microspheres using glycidyl methacrylate as monomer via photo-initiated miniemulsion polymerization," *J. Magn. Magn. Mater.*, 324(19), 3078–3082 (2012).

[6] T. D. Ngo, A. Kashani, G. Imbalzano, K. T. Nguyen, and D. Hui, "Additive manufacturing (3D printing): A review of materials, methods, applications and challenges," *Compos. Part B Eng.*, 143, 172–196 (2018).

[7] P. Parandoush, and D. Lin, "A review on additive manufacturing of polymer-fiber composites," *Compos. Struct.*, 182, 36–53 (2017).

[8] M. Shofner, K. Lozano, F. Rodríguez-Macías, and E. Barrera, "Nanofiber-reinforced polymers prepared by fused deposition modeling," *J. Appl. Polym. Sci.*, 89(11), 3081–3090 (2003).

[9] H. Chung, and S. Das, "Processing and properties of glass bead particulate-filled functionally graded nylon-11 composites produced by selective laser sintering," *Mater. Sci. Eng. A*, 437(2), 226–234 (2006).

[10] M. Nikzad, S. H. Masood, and I. Sbarski, "Thermo-mechanical properties of a highly filled polymeric composites for fused deposition modeling," *Mater. Des.*, 32(6), 3448–3456 (2011).

[11] K. Boparai, R. Singh, and H. Singh, "Comparison of tribological behaviour for nylon6-Al-Al2O3 and ABS parts fabricated by fused deposition modelling: This paper reports a low cost composite material that is more wear-resistant than conventional ABS," *Virtual Phys. Prototyp.*, 10(2), 59–66 (2015).

[12] D. Isakov, Q. Lei, F. Castles, C. Stevens, C. Grovenor, and P. Grant, "3D printed anisotropic dielectric composite with meta-material features," *Mater. Des.*, 93, 423–430 (2016).

[13] C. M. Shemelya, A. Rivera, A. T. Perez, C. Rocha, M. Liang, X. Yu, C. Kief, D. Alexander, J. Stegeman, H. Xin, R. B. Wicker, E. MacDonald, and D. A. Roberson, "Mechanical, electromagnetic, and X-ray shielding characterization of a 3D printable tungsten-polycarbonate polymer matrix composite for space-based applications," *J. Electron. Mater.*, 44, 2598–2607 (2015).

[14] R. C. D. A. G. Mota, E. O. da Silva, F. F. de Lima, L. R. de Menezes, and A. C. S. Thiele, "3D printed scaffolds as a new perspective for bone tissue regeneration: Literature review," *Mater. Sci. Appl.*, 7(8), 430–452 (2016).

[15] A. R. Torrado Perez, D. A. Roberson, and R. B. Wicker, "Fracture surface analysis of 3D-printed tensile specimens of novel ABS-based materials," *J. Fail. Anal. Prev.*, 14, 343–353 (2014).

[16] Z. Weng, Y. Zhou, W. Lin, T. Senthil, and L. Wu, "Structure-property relationship of nano enhanced stereolithography resin for desktop SLA 3D printer," *Compos. Part A Appl. Sci. Manuf.*, 88, 234–242 (2016).

[17] X. Wei, D. Li, W. Jiang, Z. Gu, X. Wang, Z. Zhang, and Z. Sun, "3D printable graphene composite," *Sci. Rep.*, 5(1), 1–7 (2015).

[18] Z. Rymansaib, P. Iravani, E. Emslie, M. Medvidović-Kosanović, M. Sak-Bosnar, R. Verdejo, and F. Marken, "All-polystyrene 3D-printed electrochemical device with embedded carbon nanofiber-graphite-polystyrene composite conductor," *Electroanalysis*, 28(7), 1517–1523 (2016).

[19] J. Hector Sandoval, and R. B. Wicker, "Functionalizing stereolithography resins: Effects of dispersed multi-walled carbon nanotubes on physical properties," *Rapid Prototyp. J.*, 12(5), 292–303 (2006).

[20] D. Lin, S. Jin, F. Zhang, C. Wang, Y. Wang, C. Zhou, and G. J. Cheng, "3D stereolithography printing of graphene oxide reinforced complex architectures," *Nanotechnology*, 26(43), 434003 (2015).

[21] D. Yugang, Z. Yuan, T. Yiping, and L. Dichen, "Nano-TiO2-modified photosensitive resin for RP," *Rapid Prototyp. J.*, 17(4), 247–252 (2011).

[22] K. Kim, W. Zhu, X. Qu, C. Aaronson, W. R. McCall, S. Chen, and D. J. Sirbuly, "3D optical printing of piezoelectric nanoparticle-polymer composite materials," *ACS Nano*, 8(10), 9799–9806 (2014).

[23] Y. Zhang, H. Li, X. Yang, T. Zhang, K. Zhu, W. Si, Z. Liu, and H. Sun, "Additive manufacturing of carbon nanotube-photopolymer composite radar absorbing materials," *Polym. Compos.*, 39(S2), E671–E676 (2018).

[24] M. He, Y. Zhao, B. Wang, Q. Xi, J. Zhou, and Z. Liang, "3D printing fabrication of amorphous thermoelectric materials with ultralow thermal conductivity," *Small*, 11(44), 5889–5894 (2015).

[25] S. R. Athreya, K. Kalaitzidou, and S. Das, "Processing and characterization of a carbon black-filled electrically conductive nylon-12 nanocomposite produced by selective laser sintering," *Mater. Sci. Eng. A*, 527(10–11), 2637–2642 (2010).

[26] H. Zheng, J. Zhang, S. Lu, G. Wang, and Z. Xu, "Effect of core-shell composite particles on the sintering behavior and properties of nano-Al2O3/polystyrene composite prepared by SLS," *Mater. Lett.*, 60(9–10), 1219–1223 (2006).

[27] H. Chung, and S. Das, "Functionally graded nylon-11/silica nanocomposites produced by selective laser sintering," *Mater. Sci. Eng. A*, 487(1–2), 251–257 (2008).

[28] F. Senatov, K. Niaza, A. Stepashkin, and S. Kaloshkin, "Low-cycle fatigue behavior of 3d-printed PLA-based porous scaffolds," *Compos. Part B Eng.*, 97, 193–200 (2016).

[29] A. B. Kutikov, A. Gurijala, and J. Song, "Rapid prototyping amphiphilic polymer/hydroxyapatite composite scaffolds with hydration-induced self-fixation behavior," *Tissue Eng. Part C Methods*, 21(3), 229–241 (2015).

[30] D. L. Naik, and R. Kiran, "On anisotropy, strain rate and size effects in VAT photopolymerization based specimens," *Addit. Manuf.*, 23, 181–196 (2018).

[31] X. Wu, Q. Lian, D. Li, X. He, X. Liu, J. Meng, and Z. Jin, "Effects of soft-start exposure on the curing characteristics and flexural strength in ceramic projection stereolithography process," *J. Eur. Ceram. Soc.*, 39(13), 3788–3796 (2019).

[32] M. Revilla-León, N. A.-H. Husain, L. Ceballos, and M. Özcan, "Flexural strength and Weibull characteristics of stereolithography additive manufactured versus milled zirconia," *J. Prosthet. Dent.*, 125(4), 685–690 (2021).

[33] E. A. Garcia, C. Ayranci, and A. J. Qureshi, "Material property-manufacturing process optimization for form 2 VAT-photo polymerization 3D Printers," *J. Manuf. Mater. Process.*, 4(1), 12 (2020).

[34] X. Wang, F. Schmidt, D. Hanaor, P. H. Kamm, S. Li, and A. Gurlo, "Additive manufacturing of ceramics from preceramic polymers: A versatile stereolithographic approach assisted by thiol-ene click chemistry," *Addit. Manuf.*, 27, 80–90 (2019).

[35] H. Liu, N. Yang, Y. Sun, L. Yang, and N. Li, "Effect of the build orientation on the mechanical behaviour of polymers by stereolithography," *IOP Conf. Ser. Mater. Sci. Eng.*, 612(3), 032166 (2019), IOP Publishing.

[36] A. Alamdari, J. Lee, M. Kim, M. O. F. Emon, A. Dhinojwala, and J.-W. Choi, "Effects of surface energy reducing agents on adhesion force in liquid bridge microstereolithography," *Addit. Manuf.*, 36, 101522 (2020).

[37] E. Aznarte Garcia, A. J. Qureshi, and C. Ayranci, "A study on material-process interaction and optimization for VAT-photopolymerization processes," *Rapid Prototyp. J.*, 24(9), 1479–1485 (2018).

[38] E. Aznarte, C. Ayranci, and A. Qureshi, "Digital light processing (DLP): Anisotropic tensile considerations," in *2017 International Solid Freeform Fabrication Symposium*, 2017: University of Texas at Austin.

[39] D. K. Patel, A. H. Sakhaei, M. Layani, B. Zhang, Q. Ge, and S. Magdassi, "Highly stretchable and UV curable elastomers for digital light processing based 3D printing," *Adv. Mater.*, 29(15), 1606000 (2017).

[40] Y.-H. Lee, J.-B. Lee, W.-Y. Maeng, Y.-H. Koh, and H.-E. Kim, "Photocurable ceramic slurry using solid camphor as novel diluent for conventional digital light processing (DLP) process," *J. Eur. Ceram. Soc.*, 39(14), 4358–4365 (2019).

[41] J. F. P. Lovo, I. L. d. Camargo, R. Erbereli, M. M. Morais, and C. A. Fortulan, "VAT photopolymerization additive manufacturing resins: Analysis and case study," *Mater. Res.*, 23, e20200010 (2020).

[42] P.-J. Wu, H. Peng, C. Li, A. Abdel-Latif, and B. J. Berron, "Adhesive stem cell coatings for enhanced retention in the heart tissue," *ACS Appl. Bio Mater.*, 3(5), 2930–2939 (2020).

[43] G. I. Peterson, J. J. Schwartz, D. Zhang, B. M. Weiss, M. A. Ganter, D. W. Storti, and A. J. Boydston, "Production of materials with spatially-controlled cross-link density via VAT photopolymerization," *ACS Appl. Mater. Interfaces*, 8(42), 29037–29043 (2016).

[44] C. J. Thrasher, J. J. Schwartz, and A. J. Boydston, "Modular elastomer photoresins for digital light processing additive manufacturing," *ACS Appl. Mater. Interfaces*, 9(45), 39708–39716 (2017).

[45] M. Monzón, Z. Ortega, A. Hernández, R. Paz, and F. Ortega, "Anisotropy of photopolymer parts made by digital light processing," *Materials*, 10(1), 64 (2017).

[46] C. He, C. Ma, X. Li, L. Yan, F. Hou, J. Liu, and A. Guo, "Polymer-derived SiOC ceramic lattice with thick struts prepared by digital light processing," *Addit. Manuf.*, 35, 101366 (2020).

[47] D. S. Kim, J. Suriboot, C.-C. Shih, A. Cwiklik, M. A. Grunlan, and B. L. Tai, "Mechanical isotropy and postcure shrinkage of polydimethylsiloxane printed with digital light processing," *Rapid Prototyp. J.*, 26(8), 1447–1452 (2020).

[48] H. Wu, P. Chen, C. Yan, C. Cai, and Y. Shi, "Four-dimensional printing of a novel acrylate-based shape memory polymer using digital light processing," *Mater. Des.*, 171, 107704 (2019).

[49] Q. Mu, L. Wang, C. K. Dunn, X. Kuang, F. Duan, Z. Zhang, H. Jerry Qi, and T. Wang, "Digital light processing 3D printing of conductive complex structures," *Addit. Manuf.*, 18, 74–83 (2017).

[50] K. Kowsari, S. Akbari, D. Wang, N. X. Fang, and Q. Ge, "High-efficiency high-resolution multimaterial fabrication for digital light processing-based three-dimensional printing," *3D Print. Addit. Manuf.*, 5(3), 185–193 (2018).

[51] S. Lantean, G. Barrera, C. F. Pirri, P. Tiberto, M. Sangermano, I. Roppolo, and G. Rizza, "3D printing of magnetoresponsive polymeric materials with tunable mechanical and magnetic properties by digital light processing," *Adv. Mater. Technol.*, 4(11), 1900505 (2019).

[52] L. Ge, L. Dong, D. Wang, Q. Ge, and G. Gu, "A digital light processing 3D printer for fast and high-precision fabrication of soft pneumatic actuators," *Sens. Actuator. A Phy.*, 273, 285–292 (2018).

[53] M. Wang, C. Xie, R. He, G. Ding, K. Zhang, G. Wang, and D. Fang, "Polymer-derived silicon nitride ceramics by digital light processing based additive manufacturing," *J. Am. Ceram. Soc.*, 102(9), 5117–5126 (2019).

[54] X. Kuang, J. Wu, K. Chen, Z. Zhao, Z. Ding, F. Hu, D. Fang, and H. Jerry Qi, "Grayscale digital light processing 3D printing for highly functionally graded materials," *Sci. Adv.*, 5(5), eaav5790 (2019).

[55] M. Liu, W. Tang, W. Duan, S. Li, R. Dou, G. Wang, B. Liu, and L. Wang, "Digital light processing of lunar regolith structures with high mechanical properties," *Ceram. Int.*, 45(5), 5829–5836 (2019).

[56] J. Schmidt, A. A. Altun, M. Schwentenwein, and P. Colombo, "Complex mullite structures fabricated via digital light processing of a preceramic polysiloxane with active alumina fillers," *J. Eur. Ceram. Soc.*, 39(4), 1336–1343 (2019).

[57] R. Dou, W. Z. Tang, L. Wang, S. Li, W. Y. Duan, M. Liu, Y. B. Zhang, and G. Wang, "Sintering of lunar regolith structures fabricated via digital light processing," *Ceram. Int.*, 45(14), 17210–17215 (2019).

[58] G. Gonzalez, A. Chiappone, K. Dietliker, C. F. Pirri, and I. Roppolo, "Fabrication and functionalization of 3D printed polydimethylsiloxane-based microfluidic devices obtained through digital light processing," *Adv. Mater. Technol.*, 5(9), 2000374 (2020).

[59] Y. Xue, L. Qi, Y. Niu, H. Huang, F. Huang, T. Si, Y. Zhao, and R. X. Xu, "Integration of electrospray and digital light processing for freeform patterning of porous microstructures," *Adv. Mater. Technol.*, 5(11), 2000578 (2020).

[60] K. Zhang, R. He, G. Ding, C. Feng, W. Song, and D. Fang, "Digital light processing of 3Y-TZP strengthened ZrO2 ceramics," *Mater. Sci. Eng. A*, 774, 138768 (2020).

[61] L. Lin, H. Wu, Y. Xu, K. Lin, W. Zou, and S. Wu, "Fabrication of dense aluminum nitride ceramics via digital light processing-based stereolithography," *Mater. Chem. Phys.*, 249, 122969 (2020).

[62] L. Wang, X. Liu, G. Wang, W. Tang, S. Li, W. Duan, and R. Dou, "Partially stabilized zirconia moulds fabricated by stereolithographic additive manufacturing via digital light processing," *Mater. Sci. Eng. A*, 770, 138537 (2020).

[63] M. Shen, W. Zhao, B. Xing, Y. Sing, S. Gao, C. Wang, and Z. Zhao, "Effects of exposure time and printing angle on the curing characteristics and flexural strength of ceramic samples fabricated via digital light processing," *Ceram. Int.*, 46(15), 24379–24384 (2020).

[64] S. G. Kim, J. E. Song, and H. R. Kim, "Development of fabrics by digital light processing three-dimensional printing technology and using a polyurethane acrylate photopolymer," *Text. Res. J.*, 90(7–8), 847–856 (2020).

[65] F. Li, X. Ji, Z. Wu, C. Qi, Q. Xian, and B. Sun, "Digital light processing 3D printing of ceramic shell for precision casting," *Mater. Lett.*, 276, 128037 (2020).

[66] B. Cao, N. Boechler, and A. J. Boydston, "Additive manufacturing with a flex activated mechanophore for nondestructive assessment of mechanochemical reactivity in complex object geometries," *Polymer*, 152, 4–8 (2018).

[67] M. Shen, W. Qin, B. Xing, W. Zhao, S. Gao, Y. Sun, T. Jiao, and Z. Zhao, "Mechanical properties of 3D printed ceramic cellular materials with triply periodic minimal surface architectures," *J. Eur. Ceram. Soc.*, 41(2), 1481–1489 (2021).

[68] B. Steyrer, P. Neubauer, R. Liska, and J. Stampfl, "Visible light photoinitiator for 3D-printing of tough methacrylate resins," *Materials*, 10(12), 1445 (2017).

[69] T. Hada, M. Kanazawa, M. Iwaki, T. Arakida, and S. Minakuchi, "Effect of printing direction on stress distortion of three-dimensional printed dentures using stereolithography technology," *J. Mech. Behav. Biomed. Mater.*, 110, 103949 (2020).

[70] N. A. Chartrain, M. Vratsanos, D. T. Han, J. M. Sirrine, A. Pekkanen, T. E. Long, A. R. Whittington, and C. B. Williams, "Microstereolithography of tissue scaffolds using a biodegradable photocurable polyester," in *2016 International Solid Freeform Fabrication Symposium*, 2016: University of Texas at Austin.

[71] S. D. Silbert, P. Simpson, R. Setien, M. Holthaus, J. La Scala, C. A. Ulven, and D. C. Webster, "Exploration of bio-based functionalized sucrose ester resins for additive manufacturing via stereolithography," *ACS Appl. Polym. Mater.*, 2(7), 2910–2918 (2020).

[72] R. B. Osman, A. J. van der Veen, D. Huiberts, D. Wismeijer, and N. Alharbi, "3D-printing zirconia implants; a dream or a reality? An in-vitro study evaluating the dimensional accuracy, surface topography and mechanical properties of printed zirconia implant and discs," *J. Mech. Behav. Biomed. Mater.*, 75, 521–528 (2017).

[73] M. Maturi, C. Pulignani, E. Locatelli, V. V. Buratti, S. Tortorella, L. Sambria, and M. C. Franchini, "Phosphorescent bio-based resin for digital light processing (DLP) 3D-printing," *Green Chem.*, 22(18), 6212–6224 (2020).

[74] S. H. Kim, Y. K. Yeon, J. M. Lee, J. R. Chao, Y. J. Lee, Y. B. Seo, M. T. Sultan, O. J. Lee, J. S. Lee, S.-I. Yoon, I.-S. Hong, G. Khang, S. J. Lee, J. J. Yoo, and C. H. Park, "Precisely printable and biocompatible silk fibroin bioink for digital light processing 3D printing," *Nat. Commun.*, 9(1), 1620 (2018).

[75] G. González, D. Baruffaldi, C. Martinengo, A. Angelini, A. Chiappone, I. Roppolo, C. F. Pirri, and F. Frascella, "Materials testing for the development of biocompatible devices through VAT-polymerization 3d printing," *Nanomaterials*, 10(9), 1788 (2020).

[76] H. Chen, S.-Y. Lee, and Y.-M. Lin, "Synthesis and formulation of PCL-based urethane acrylates for DLP 3D printers," *Polymers*, 12(7), 1500 (2020).

[77] C. Paredes, F. J. Martinez-Vazquez, H. Elsayed, P. Colombo, A. Pajares, and P. Miranda, "Evaluation of direct light processing for the fabrication of bioactive ceramic scaffolds: Effect of pore/strut size on manufacturability and mechanical performance," *J. Eur. Ceram. Soc.*, 41(1), 892–900 (2021).

[78] J. Schmidt, H. Elsayed, E. Bernardo, and P. Colombo, "Digital light processing of wollastonite-diopside glass-ceramic complex structures," *J. Eur. Ceram. Soc.*, 38(13), 4580–4584 (2018).

[79] D. Xue, J. Zhang, Y. Wang, and D. Mei, "Digital light processing-based 3D printing of cell-seeding hydrogel scaffolds with regionally varied stiffness," *ACS Biomater. Sci. Eng.*, 5(9), 4825–4833 (2019).

[80] M. Krkobabić, D. Medarević, S. Cvijić, B. Grujić, and S. Ibrić, "Hydrophilic excipients in digital light processing (DLP) printing of sustained release tablets: Impact on internal structure and drug dissolution rate," *Int. J. Pharm.*, 572, 118790 (2019).

[81] Y. Zeng, Y. Yan, H. Yan, C. Liu, P. Li, P. Dong, Y. Zhao, and J. Chen, "3D printing of hydroxyapatite scaffolds with good mechanical and biocompatible properties by digital light processing," *J. Mater. Sci.*, 53(9), 6291–6301 (2018).

[82] Z. Liu, H. Liang, T. Shi, D. Xie, R. Chen, X. Han, L. Shen, C. Wang, and Z. Tian, "Additive manufacturing of hydroxyapatite bone scaffolds via digital light processing and in vitro compatibility," *Ceram. Int.*, 45(8), 11079–11086 (2019).

[83] J.-H. Kim, W.-Y. Maeng, Y.-H. Koh, and H.-E. Kim, "Digital light processing of zirconia prostheses with high strength and translucency for dental applications," *Ceram. Int.*, 46(18), 28211–28218 (2020).

[84] Y. Wei, D. Zhao, Q. Cao, J. Wang, Y. Wu, B. Yuan, X. Li, X. Chen, Y. Zhou, X. Yang, X. Zhu, C. Tu, and X. Zhang, "Stereolithography-based additive manufacturing of high-performance osteoinductive calcium phosphate ceramics by a digital light-processing system," *ACS Biomater. Sci. Eng.*, 6(3), 1787–1797 (2020).

[85] A. B. Saed, A. H. Behravesh, S. Hasannia, S. A. A. Ardebili, B. Akhoundi, and M. Pourghayoumi, "Functionalized poly l-lactic acid synthesis and optimization of process parameters for 3D printing of porous scaffolds via digital light processing (DLP) method," *J. Manuf. Process.*, 56, 550–561 (2020).

[86] C. Feng, K. Zhang, R. He, G. Ding, M. Xia, X. Jin, and C. Xie, "Additive manufacturing of hydroxyapatite bioceramic scaffolds: Dispersion, digital light processing, sintering, mechanical properties, and biocompatibility," *J. Adv. Ceram.*, 9, 360–373 (2020).

[87] Y. Luo, G. Le Fer, D. Dean, and M. L. Becker, "3D printing of poly (propylene fumarate) oligomers: Evaluation of resin viscosity, printing characteristics and mechanical properties," *Biomacromolecules*, 20(4), 1699–1708 (2019).

Index

Note: **Bold** page numbers refer to tables and *italic* page numbers refer to figures.

ABD matrices, writing program to find 36–42, *37–41*
absorbing energy, mechanisms for 87–88
additive manufacturing (AM) 136–137, *138*, *139*, 151
  biomedical applications 183–184, **184–187**
  comparison of 147, **148**, *149*
  composites, filler materials **160–161**, 160–162
  in composites, general applications 162–163
  extrusion-based AM process 142–143
  manufactured composites with **176**
  mechanical applications 174
  powder-based AM process 144–146
  techniques with improved physical characteristics **175**
  types 141–142, *142*
additive manufacturing polymer composites, design for 151
  design freedom 165
  performance 153
  process 151
  properties 153
  PSPP framework 151, *152*
  structure 153
additive shaping process 139, *140*, *142*
adjustable structural performance and CTE 168–169
aerogel microlattice, graphene 167
Algorithms of Ingenious Problem Solving (ARIZ) 118
aluminum 1
AM *see* additive manufacturing (AM)
ASTM D3039 and D790 methods 155
ASTM standard 127, 137, 149n1
automotive businesses 113
automotive industry, composite materials in 125
automotive interiors 125
axial loading, FEM simulation models 62

basalt fibers (BFs) 98
  characteristics, applications 98–100
  made by melt-blown process 100–101
  mat-reinforced hybrid thermosets 106–107
basalt rock 98, 100
bend curved beams 55–61, *57*

bending
  FEM simulation models 62–63
  of plate 70
Bernoulli-Navier hypothesis 48
BFs *see* basalt fibers (BFs)
bi-axial loading 73
bidirectional fibers 8
biomedical applications in additive manufacturing 183–184, **184–187**
Boeing models 162
bonnet 127
boron filaments 94
brainstorming method 120
brittle materials 88
broken fiber 1, *2*
buckling of plate 70–71
  bi-axial loading 73
  clamped on all sides 72
  fixed un-loaded edges 73
  simply supported on all edges 71–72
  single free edge 72–73
buckling of stiffened panel 77–79

carbon fiber-reinforced polymer (CFRP) 157, 162
carbon fibers 156, 166
carbon nanofibers 144
carbon nanotubes (CNTs) 174
CF-reinforced epoxy matrix (EP) composites 102
CF-reinforced PP composites 104
CFRP *see* carbon fiber-reinforced polymer (CFRP)
Charpy impact machine 90
Charpy impact test 91
classical laminated plate theory (CLPT) 25, 26
classical laminate theory (CLT) 25
classical plate theory (CPT) 25–30
close-section stiffeners 76
CLPT *see* classical laminated plate theory (CLPT)
CNTs *see* carbon nanotubes (CNTs)
coefficient of thermal expansion (CTE) 165
  adjustable structural performance and 168–169
component materials properties 6

composite
  fracture process in 87
  impact behavior of 90
  piezoelectric 45
composite beams, conceptual design 37, *37*,
    48, 64
  behavior of column 68–69
  buckling design 66–67
  deflection design 64, *65*
  Euler buckling 68, **68**
  local buckling 68–69
  material error 69
  mode interaction 69
  strength design 66
composite elements, stress-strain relations for
    on-axis and off-axis 48–50
composite materials
  in automotive industry 125
  conventional failure theories for 46–47
  failure analysis in 45–46
  micromechanics techniques for 7
  modulus of 11
  multilayer 125–126, *126*
  S-N curves of 86
  Young's modulus of 11, *12*
computer aided design software (CAD) 137
concept design 115–116
construction-related industries 114
consumers 112
continuous carbon fibers 166
continuous-fiber composites 5
continuous-fiber-reinforced composites 82
Continuous Fiber Reinforced Thermoplastic
    Composite (CFRTPCs) 154
Contradiction Engineering with 40 Ingenious
    Principles 118
conventional failure theories, for composite
    materials 46–47
coupling agent 94
CPT *see* classical plate theory (CPT)
crack
  arrest 83–84
  branching 83–84, 89
  cross-ply 84, 85
  initiation 83–84
  matrix deformation and 89
cross-ply cracks 84, 85
crushed basalt rock 98
CTE *see* coefficient of thermal expansion
    (CTE)
curved beams, bend 55–61, *57*
cylinder-shaped shells 30

dampers, piezoelectric passive 45
debonding fractures 90
deflection design, composite beams 64, *65*

degree of freedom (DOF) *131*, 132
delamination 91
  cracks 84
deposition of materials, layer-by-layer 165
design for excellence (DfX) 117
design for sustainability (DfS) 116, 121
digital light processing (DLP) 176
  polymers produced via DLP process for
    biomedical applications **185–187**
  polymers produced via DLP process for
    mechanical applications **178–182**
direct ink writing (DIW) 143, *143*
  additive manufacturing of silica aerogel
    parts 167
direct light process (DLP) 146
DIW *see* direct ink writing (DIW)
DMLS 162
dual-material additive manufacturing process,
    polymer composites 168

electrically conductive inks 164
E_1 finding 11–13
E_2 finding 13–16
E-glass fibers 94
elastic behavior, nonlinear 1–2
elastic constants 53
elastic materials, linear 2
elastic moduli 85
electrical conduction, polymer composites
    163–164
electrically conductive PMCs (eMCS) 143
electromagnetic metamaterials 164
electromagnetic shielding, polymer composites
    164–165
electronic components, NFC applications in
    121
eMCS *see* electrically conductive PMCs
    (eMCS)
engine cover 127
environmental-interaction effects 92
epoxies 89
epoxy resin (EP) 100
Euler-Bernoulli beam theory 25
extrusion-based AM process 142–143

failure analysis in composite materials 45–46
failure models 87–88
fatigue analysis 82
fatigue damage 83
  crack arrest and crack branching 84
  damage/crack initiation 83–84
  empirical relations for 86–87
  final fracture 84
  influence of damage on properties 85–86
fatigue life 85
  empirical relations for 86–87

F42 committee 149n2
FDM *see* fused deposition modeling (FDM)
FEM simulation models 61–63
    axial loading 62
    bending 62–63
FFF *see* fused filament fabrication (FFF)
fiber 8, 90; *see also specific types*
    bidirectional 8
    breakage 88
    debonding 89–90
    high-quality 98
    natural (*see* natural fiber)
    quantity of 8
    stiff 1
    stress-rupture of 93–94
    synthetic 158
    unidirectional *1, 9*
    winding resin-coated 3
    woven *1*
fiber-reinforced composite materials 2,
        7, 122
fiber-reinforced laminate, load in *12*
fiber-reinforced polymers (FRPs) 98
    composites 108
    impact behavior 90
fiber-spinning process 101
filler materials in additive manufacturing
        composites **160–161**, 160–162
forcing adoption through function 114
formative shaping 139, *140, 141*
fracture energy 89
fracture mechanics, linear elastic 104
fracture process in composites 87
frames 74
FRPs *see* fiber-reinforced polymers (FRPs)
fused deposition modeling (FDM) 154, 174, **175**
fused filament fabrication (FFF) 142

gallery method 120
Galvano scanners 146
G_{12} finding 18–20, *19, 20*
glass fiber 94, 156
glass-fiber-reinforced plastic 90
glass-reinforced polypropylene 85–86
graphene aerogel microlattice 167
graphene-based materials, polymer
        composites 166

Halpin-Tsai equations 7
high-density polyethylene (HDPE) matrix 100
high-quality fibers 98
high-speed photography 90
high-strain rate 87
high-strength fiber 5
Hoffman failure theory 47, *47*
hood panel 127

Hooke's law 53
    strain symmetry 54–55
HXs (composite heat exchangers) 163
hybrid composites 101–103
    reinforcement content and composition
        of **104**
    thermoplastic 104–105
    thermoset 105–106
hybrid effect 103
hybrid fiber mat-reinforced hybrid
        thermosets 107
hybrid fiber-reinforced composites, mechanical
        characteristics **105**
hybrid thermosets
    basalt fiber mat-reinforced 106–107
    hybrid fiber mat-reinforced 107, *107*
hygrothermal stress/strain 42–45, *45*

impact energy 92
impact loads 87
impact properties, effect of materials and
        testing variables on 90–92
industrial design (ID) 114
interface strength 91
IPN-structured BF mat-reinforced VE/EP
        resins, mechanical characteristics
        of **106**

Kelly's equation 89
Kevlar fiber *156*
Kevlar fiber-reinforced composites 155
Kirchhoff-Love theory *see* classical plate
        theory (CPT)
Kirchhoff's hypotheses 25, 30

lamina 8, *9*
    definition 1–2, 5
    stress-strain of 26
    with unidirectional fibers *9*
laminate
    bending moduli 71
    definition *1*, 2–3, *4*
    variation of stress and strain in 27,
        27–30, *28*
layer-by-layer deposition of materials 165
linear elastic fracture mechanics 104
linear elastic materials 2
linear stress-strain relationship, stiffness
        coefficients for 53
load transfer mechanism 1
longitudinal-ply cracks 84
longitudinal-ply fractures 84
low-pressure moldable SMC 125
LPMC (low-pressure molding compound)
        125, 127
lumped model 127

macromechanics 6
material properties, effects of 83
materials approach
    impact behavior 87
    mechanics of 6, 9
    selection 114–115
    strength 82
    stress-strain curve 85
    using strength of 55–61
matrix 8
    assessment 116
    deformation and cracking 89
maximum strain failure criterion 46
maximum stress failure criterion 46
Meanings of Material (MoM) model 115
mechanical characteristics, polymer composites
        153, **154**
    comparison 157
    energy absorption and damping mechanism
        155–157
    flexural characteristics 154–155, *156*
    tailoring energy absorption and
        damping 157
    tensile and flexural properties 154
    tensile characteristics 154
mechanics of materials approach 6, 9
melt-blown process, basalt fibers made by
        100–101
metals 64
metamaterials 167
    electromagnetic 164
microchannels 163
microlattice, graphene aerogel 167
micromechanical stiffness technique 8
micromechanical theories of stiffness 7
micromechanics 5, 6
    forms 10
    techniques for composite materials 7
MicroPerforated Panel (MPP) 158
    sound absorption coefficient (SAC) on 159
Mitsubishi electric Melta RV-2F robotic
        manipulator 143
MJF *see* multiple jetting fusion (MJF)
modal analysis, case study 127–132, **128**,
        131, *131*
morphological chart 119–120
MPP *see* MicroPerforated Panel (MPP)
multilayer composite materials 125–126, *126*
multi-material additive manufacturing
        advancement 168
multiple jetting fusion (MJF) *145*, 145–146

nanofibers, carbon 144
natural fiber composites (NFC) 112–113, 122
    adoption of 114

characteristics 112–113
    design 114
natural fiber composites (NFC) applications
        120–121
    in electronic components 121
    in packaging 121–122
    in sports equipment 122
natural fiber polymer composites (NFPC)
        112, 113
natural fiber polymer composites (NFPC)
        applications 113
    automotive businesses 113
    construction-related industries 114
    forcing adoption through function 114
natural fiber-reinforced polymer
        composites 113
Newton's second law 133
NFPC *see* natural fiber polymer composites
        (NFPC)
nonlinear elastic behavior 1–2

off-axis composite elements, stress-strain
        relations for 48–50
off-axis system 50–53, *52*
on-axis composite elements, stress-strain
        relations for 48–50
organic additive fillers 162

packaging, NFC applications in 121–122
partly constrained layer (PCL) damping
        treatments 126
Paugh method 116
PDS *see* product design standards (PDS)
petroleum-based plastics 121
phonon propagation and thermal
        conductivity 166
photopolymerization-based process *146*,
        146–147
piezoelectric composites 45
piezoelectric passive dampers 45
plastics, petroleum-based 121
PMC *see* polymer matrix composites (PMC)
Poisson effects 77
Poisson's ratio 17, 20
polyamides 145
polycarbonate (PC) 100
polyesters 89
polylactic acid/polyhydroxyalkanoates-wood
        fibers (PLA/PHA-WF) composite 158
polymer composites 163, 174
    acoustic characteristics driven design *158*,
        158–160, **159**
    adjustable structural performance and CTE
        168–169
    continuous carbon fibers 166

DIW additive manufacturing of silica
    aerogel parts 167
dual-material additive manufacturing
    process 168
electrical and electromagnetic
    characteristics 163–165
electrical conduction 163–164
electromagnetic shielding 164–165
enhanced thermal insulation 167–168
enhancing thermal conductivity via
    fillers 166
fiber-reinforced 108
graphene aerogel microlattice 167
graphene-based materials 166
multi-material additive manufacturing
    advancement 168
phonon propagation and thermal
    conductivity 166
produced via SLA process for mechanical
    applications 177
reinforcing thermal conduction
    characteristics 166–167
stretch-dominated bi-material unit cell 168
thermal conduction and expansion 165–169
thermal conductivity 166
thermal insulative 167
three-phase optimization topology 168
polymer composites, mechanical characteristics
    153, 154
comparison 157
energy absorption and damping mechanism
    155–157
flexural characteristics 154–155, 156
tailoring energy absorption and
    damping 157
tensile and flexural properties 154
tensile characteristics 154
polymer matrix composites (PMC) 143
polymers 112, 147, 174
    produced via DLP process for biomedical
        applications 185–187
    produced via DLP process for mechanical
        applications 178–182
    produced via SLA process for biomedical
        applications 184
    produced via SLA process for mechanical
        applications 177
polypropylene (PP) 100
polystyrene (PS) composite plates 100
powder-based AM process 144–146
power law 87
process-structure-properties-performance
    (PSPP) framework 151, 152
product designers 116
product design specifications (PDS) 116
product design standards (PDS) 116

product development
    incorporating sustainable design with other
        concurrent engineering processes
        during 116–117, 118
property prediction 103

QFD (quality function deployment/product
    planning matrix) 119

rapid prototyping (RP) 136, 149n1
real design power 6
reinforcing fibers, properties of 101
representative volume element 8, 9, 11
    shear-loaded 19
resin, UV light-introduced polymerization
    of 147
resistance of materials 6
rock wool 98
RP see Rapid Prototyping (RP)

Scott Bader Company 125
selective laser sintering (SLS) 144,
    144–145, 168
shaping types 139, 140
shear deformation theory 28
shear-loaded representative volume element 19
sheet molding compound (SMC) 125
silicate glasses 93
single free edge, buckling of plate 72–73
SLM 162
SLS see selective laser sintering (SLS)
SMC see sheet molding compound (SMC)
solution-precipitation approach 163
sound absorption coefficient (SAC) on MPP 159
sports equipment, NFC applications in 122
stretch-dominated bi-material unit cell 168
static fatigue 93–94
stereolithography (SLA) 146, 147, 174, 175
    polymers produced via SLA process for
        biomedical applications 184
    polymers produced via SLA process for
        mechanical applications 177
stiffened panels 73–74
stiffened panels under loads with bending 74
    buckling of stiffened panel 77–79
    crippling and post-buckling 79–81
    design of skin 74–75
    design of stringer 75–76
    under in-plane loads 76
    stiffener and skin strength 76–77
stiff fibers 1
stiffness
    coefficients for linear stress-strain
        relationship 53
    matrix 23
    micromechanical theories of 7

strain
  failure criterion 46
  stress and 23–25
  symmetry 54–55
strength of materials 6
stress and strain 23–25
  finding 33–36, *35*
  variation, in laminate *27*, 27–30, *28*
stress corrosion, features of 93
stress failure criterion 46
stress-rupture
  of fibers 93–94
  test 94
stress-shear strain behavior 18
stress-strain
  behaviour *3*
  hygrothermal 42–45, *45*
  of lamina 26
stress-strain relations for on-axis and off-axis
  composite elements 48–50
  off-axis system 50–53, *52*
stress symmetry 53
subtractive shaping 139, *140*, *141*
Su-field modeling 118
surface roughness 163
synthetic fiber 158

Technology Trend Prediction 118
tensile test 94
theory of inventive problem-solving (TRIZ)
  117–118, *119*
  brainstorming method 120
  chart for morphology 119–120
  customer voice 118–119
  gallery method 120
  increasing search space 120
thermal conductivity, polymer
  composites 166
thermal insulation, polymer composites
  167–168
thermoplastics 112
  hybrid composites 104–105

matrix materials 112
  for SLS 144, *144*
thermoset hybrid composites 105–106
thin plate 30–31, *31*
3D scanners 137
3D computer-aided design (3D CAD) 137
3D printing *see* additive manufacturing (AM)
three-phase optimization topology 168
traditional impact-testing instruments 90
transformation matrix 22–23
transverse reinforcements 74
transverse shear deformation, effects of 71
TRIZ tool 118
Tsai-Hill failure theory 46
two- or three-layer damping treatment's modal
  analysis
  constrained layer damping treatment
  133–134
  unconstrained layer damping treatment
  132–133

unidirectional fibers *1*
  lamina with *9*
universal testing machine (UTM) 127
unsaturated polyester (UP) 125
UP *see* unsaturated polyester (UP)
UTM *see* universal testing machine (UTM)
UV laser light 146
UV light-introduced polymerization of
  resin 147

vat polymerization (VP) 146–147
V_{12} finding *17*, 17–18, *18*
vibration management components 125–126
vinylester (VE) 106
voice of the customer (VOC) 118

whisker-reinforced composite materials 1
winding resin-coated fibers 3
woven fibers *1*

Young's modulus of composite material 11, *12*

Printed in the United States
by Baker & Taylor Publisher Services

Printed in the United States
by Baker & Taylor Publisher Services